People and place

PEARSON EDUCATION

We work with leading authors to develop the strongest educational materials in geography, bringing cutting-edge thinking and best learning practice to a global market.

Under a range of well-known imprints, including Prentice Hall, we craft high quality print and electronic publications which help readers to understand and apply their content, whether studying or at work.

To find out more about the complete range of our publishing please visit us on the World Wide Web at: www.pearsoneduc.com

People and place

The extraordinary geographies of everyday life

Lewis Holloway and
Phil Hubbard

An imprint of Pearson Education

Harlow, England · London · New York · Reading, Massachusetts · San Francisco · Toronto · Don Mills, Ontario · Sydney
Tokyo · Singapore · Hong Kong · Seoul · Taipei · Cape Town · Madrid · Mexico City · Amsterdam · Munich · Paris · Milan

Pearson Education Limited
Edinburgh Gate
Harlow
Essex CM20 2JE
England

and Associated Companies around the world.

Visit us on the World Wide Web at:
www.pearsoneduc.com

First published 2001

ISBN 0 582 38212 2

British Library Cataloguing-in-Publication Data
A catalogue record for this book can be obtained from the British Library

Library of Congress Cataloging-in-Publication Data
A catalog record for this book can be obtained from the Library of Congress

10 9 8 7 6 5 4 3 2 1
05 04 03 02 01 00

Typeset in 10/12.5 Sabon by 42

Printed in Great Britain by Henry Ling Ltd, at the Dorset Press,
Dorchester, Dorset.

contents

list of figures

list of tables

preface: read this!

Although this book will hopefully be read by many different people, for many different reasons, its original intention was simply to focus on the very different (and often conflicting) ways that Anglo-American human geographers have sought to imagine and explore relationships between people and place. Straight away, when writing a geographical textbook of this type we are faced with the problem of how to avoid giving the impression that geography is an inherently ordered discipline where it is possible to pigeonhole particular ways of 'doing' geography into easy categories. Nonetheless, we have divided the text into a number of chapters which examine how geographers working within particular research traditions have sought to study the relationships between people and place. This emphasizes the differences between specific approaches and stresses their distinctive features. However, while to some extent this glosses over the messiness of geographical histories (Livingstone, 1992), we have tried not to lose a sense of the ways in which the different research traditions have co-existed and taken form partly in response to each other. This implies that when we read about any one approach to geography, alternative approaches are always in some way present as they are explicitly or implicitly questioned or criticized by the approach in question. To a certain extent, while we are wary of using militaristic metaphors, the presentation of a selection of geographical research traditions might be understood as a series of skirmishes between competing ways of understanding geographical relationships. Such a contest to see what theory best allows us to understand the relationships between people and place might imply that geographers have tended to entrench themselves in fixed theoretical positions, taking potshots at opposing theories and ducking the returning fire while composing defences of their own geographical explanations. While this has sometimes seemed to be the case, it is perhaps better to understand geographical debate as a process in which the positions taken by individual geographers are constantly changing and where the terms of reference used by geographers to discuss the world from a geographical perspective are in a state of continual flux. Thus, we suggest that it is possible to explore intertwining strands of geographical thought, rather than maintain distinctions between entirely separate schools of thought. This raises some important questions about our ability to talk about making 'progress' in geographical enquiry, or our ability to assert that a particular theoretical tradition constitutes geographical 'truth'. Throughout the book, we avoid forcing choices between approaches, concluding with an understanding that each of the many different theoretical perspectives offers us the chance to learn something about the relationships between people and place.

As you read this text, it is important for you to realize that it is not (and, indeed, could never be) a definitive account of the ways that geographers have sought to explore people–place relations. Like other geography texts, it is an account that inevitably ends up ignoring or downplaying the contribution of particular intellectual traditions (and particular individuals) while highlighting others. Paul Feyerabend (1978) has stated that there is no idea which is not capable of improving our knowledge, and it is clear to us that geographers should always be wary of excluding insights and ideas developed by those writing in other academic disciplines, other languages and even beyond academia all together. Yet we also acknowledge that we are writing for an audience situated largely within institutions of higher education and having certain expectations of the type and range of work that we will draw on in a text like this. As such, the majority of the chapters that follow focus mainly on key ideas (highlighted in **bold** type) which, for better or worse, have become widely discussed and debated within 'mainstream' Anglo-American academic geography. The geographer David Sibley (1995) has emphasized that what is regarded as mainstream is defined by certain powerful geographers, and has marginalized and ignored the views of those whose ideas about the world do not conform with the assumptions that they hold. Rather than accepting, or even celebrating, a plurality of different ideas, geography as a discipline has tended to be characterized by its support for (and protection of) the ideas propagated by its most influential and powerful practitioners at any given time. This can lead to the illusion of a unified and strong discipline which is confident about its role and its ability to advance knowledge. Such an illusion might act to hide the constant undercurrents and strategies of academic debate, not to mention intradisciplinary power struggles between different individuals, university departments and theoretical approaches. And more recently, geography has found itself becoming more open and responsive to ideas generated outside of its disciplinary boundaries and from individuals and groups prepared to challenge dominant understandings of how geography should be 'done'. Geographers have begun to take into account 'other' voices: those voices which have been overtly or covertly silenced in an academic world which has so often been dominated by middle-class, white, ablebodied, heterosexual men. Thus, the voices and geographical experiences of (for example) women and people from different ethnic backgrounds and social classes have increasingly been influential in the development of geographical thought. In response, the discipline has become rather more 'undisciplined' (and, perhaps, more *self*-critical and humble) in its engagement with ideas which question notions of truth and the possibility of complete explanations of social, cultural and economic phenomena. Such an engagement with what might be very broadly labelled 'postmodern' theory (something examined towards the end of this book) disrupts the coherent narratives and explanations of much geographical work, recognizing the importance of change, difference, diversity and fragmentation in our experience of a rude and ill-mannered (rather than an always polite and orderly) world. To talk about 'geography' in isolation from other areas of academic enquiry is, therefore, futile, and we would encourage you to gain a sense of exploratory 'indiscipline' at the same time that you begin to recognize the value of a geographical sensitivity in trying to make sense of the world. Much of what you will find in this book is sociological, cultural, historical, poetic, philosophical and psychological. As we demonstrate in the book, these are things which are always already geographical. We hope that this is not just an eclectic and interesting mix, but that it contributes to a questioning of strict disciplinary boundaries.

While we don't want to tell you how to read this book (we would prefer that you make your own unique reading of it, which might be entirely different to that of your peers – and probably quite different to the reading we imagined you might make!), there are some important points we would like to make at this stage, before you explore the book and the ideas it con(s)t(r)ains and (hopefully) sparks off in your own imagination. Firstly, it is not necessary to read the book in the order in which it is presented, although we certainly recommend that you read Chapter 1, '. . . Arrivals', before anything else. Each of the nine main chapters (Chapters 2–10) is designed to 'stand alone', but, importantly, each chapter is deliberately left open to interference (i.e. ideas, interjections and interruptions) from other chapters. As mentioned above, in exploring any one approach, other approaches are always present, either explicitly or lurking as shadows in the background. In order to facilitate such interference, and to help you to navigate your way around the book, you will find numerous cross-references between the content of specific chapters, which you can follow up immediately or during later reading.

However, not all the interference experienced in this book is internal, between chapters. We have also constantly demonstrated the supportive and critical interference between our chapters and much wider sets of academic and non-academic literatures. Thus, you will find suggestions for further reading, quotations from other authors, and references indicating the sources of the concepts discussed during the various chapters of the book.

Finally, we introduce another source of external interference – you the reader – suggesting a number of 'exercises' which encourage you to engage directly with the geographical concepts under discussion, and which should lead to the production of your own new ideas based on your own experience: something which only you can bring to the book.

acknowledgements

In an introductory text such as this, it is inevitable that we have drawn on the ideas and arguments of a huge range of individuals. While we reference many of these in the text, we should perhaps also acknowledge the influence here of some of those who remain uncited, especially those who inspired us to become geographers in the first instance. We would particularly like to thank those teachers and lecturers whose passion for geography rubbed off on us, as well as the students who continue to maintain our enthusiasm for the subject. Here, a particular debt is owed to those human geography undergraduates and postgraduates at Coventry University who have provided sometimes critical but always useful feedback on many of the ideas worked through in this book. Additionally, we have both been lucky enough to be surrounded by generous and supportive colleagues while working on the book, and many of these have contributed to its writing by cajoling us to explore particular lines of flight. Here, we particularly want to thank colleagues at Loughborough and Coventry, past and present, including Rosie Cox, Marcus Doel, Phil Dunham, Sarah Holloway, Moya Kneafsey, Brian Ilbery, Martin Phillips, Jon Stobart and Adrian Wood (among others). More widely, this text has benefited from criticisms and suggestions by Tim Hall, Rob Kitchin and Keith Lilley. Phil would also like to thank Kieran for the photos, Cathy for her love and support, and Kid Loco for calming him down when things got a bit much; Lewis would like to thank Lucy for TLC and puppy love, mum and dad for love and encouragement over the years and the smallholders who allowed him to let off steam in their mud. Finally, we are indebted to Matthew Smith, whose constant reassurance and encouragement has made the publication of this book a relatively painless process.

The publishers are grateful to the following for permission to reproduce copyright material:

'Prescott to unveil new national parks' by Lucy Ward and Keith Harper, 29 September 1999, an extract from 'The trouble with Boothby Graffoe' by J. Cunningham, 12 August 1997 and an extract from a report dated 5 January 1998 © The Guardian; 'Fern Hill' by Dylan Thomas, from *The Poems of Dylan Thomas*, copyright © 1945 by the Trustees for the Copyrights of Dylan Thomas. Reprinted by permission of New Directions Publishing Group (US rights); 'Fern Hill' by Dylan Thomas, from *Collected Poems*, published by J.M. Dent, reprinted by permission of David Higham Associates Limited (world rights).

Figure 1.1 photograph by Roger Bamber; Figure 2.1 from *Globalization and World Cities*, J. Short and Y. M. Kim, 1999, Figure 3.5 from *Making Sense of Place:*

children's understanding of large-scale environments, H. Matthews, 1992, Figure 3.8 from *Urban Living: the individual in the city*, D. J. Walmsley, 1988, Figures 7.11 and 7.12 from *Mapping: ways of representing the world*, D. Dorling and D. Fairbairn, 1997, all reproduced with permission from Pearson Education Limited; Tables 2.2 and 2.3 from *Social Trends 1998*, National Statistics © Crown Copyright 2000; Figure 2.2 photograph supplied by Modus Publicity, on behalf of Benetton; Figure 2.4 from *Times, Spaces and Places: a chronogeographic perspective*, D. Parkes and N. Thrift, 1980, reproduced by permission of John Wiley & Sons, Ltd; Figure 2.5 reprinted from *Health and Place*, Vol 2, Wilton, R. 'Diminished Worlds? The geography of everyday life with HIV/AIDS', p.77, Copyright (1996), with permission from Elsevier Science; Figure 2.6 from *Lost Words and Lost Worlds: Modernity and the Language of Everyday Life in Nineteenth-Century Stockholm*, Allan Pred, 1990, reproduced by permission of Cambridge University Press; Figure 4.3 © Tony Stone Images; Figure 4.5 *Crows in the wheatfields* reproduced with permission from the Van Gogh Museum (Vincent van Gogh Foundation); Figure 5.1 Digital Imagery © PhotoDisc, Inc.; Figures 5.2 and 5.3 from 'The Black Inner City as Frontier Outpost: Images and behavior of a Philadelphia neighborhood' D. Ley and R. Cybriwsky, 1974, monography No.7, pages 131 and 132, Association of American Geographers; Figures 5.4 and 8.6 © Punch Limited; Figure 5.5 from 'The English and their Englishness: a curiously mysterious, elusive and little understood people' in *Scottish Geographical Magazine*, 107, pp. 146-161, Royal Scottish Geographical Society; Figure 6.2, 'Stockbroker's Tudor' from *Pillar to Post* by Osbert Lancaster reproduced with permission from John Murray (Publishers) Ltd; Figure 6.4 from *Maps and Meaning*, P. Jackson, 1989, Routledge; Figure 6.5 front cover illustration from *Personal Narrative of a Pilgrimage to Al-Madinah and Meccah* Volume 1, Sir Richard F Burton, 1964, Dover Publications, Inc; Figure 7.1 *The Treachery of Images*, 1929 by René Magritte, 1929 © ADAGP, Paris and DACS, London 2000; Figure 7.2 *Mr and Mrs Andrews* by Thomas Gainsborough © National Gallery, London; Figure 7.3 from *Grundy's English Views* (1857) reproduced with permission from Hulton Getty; Figure 7.4 (a) reproduced courtesy of J. Barbour & Sons Ltd; Figure 7.4 (b) courtesy of Stanley Cookers (GB) Limited; Figure 7.6 courtesy of Allen & Page Specialist Feeds; Figure 7.7 © Peter Arkell/Impact Photos; Figure 7.8 reproduced with permission from Aerofilms Ltd; Figure 7.13 Map reproduced from Ordnance Survey mapping with the permission of the Controller of Her Majesty's Stationery Office, © Crown Copyright; Figure 8.3 © Clive Shirley/Impact Photos; Figure 9.2 © Simon Shepheard/Impact Photos; Figure 9.3 reproduced with permission from The Ramblers' Association; Figure 9.4 photograph by Roy Riley © Times Newspapers Limited; Figure 10.1 from *Place/Culture/Representation*, J. Duncan and D. Ley, 1993, Routledge.

Whilst every effort has been made to trace the owners of copyright material, in a few cases this has proved impossible and we take this opportunity to offer our apologies to any copyright holders whose rights we may have unwittingly infringed.

one ... Arrivals

This chapter covers:

1.1 Introduction

It has become commonplace for geographers to begin their texts by making reference to the fact that 'geography matters'. Typically, this is to make a claim that geography can illuminate debates about a range of social, economic and political issues, and to stress that the discipline is one which is relevant, important and up-to-date. This is often shown with reference to matters which are clearly of global importance – environmental pollution, the debt crisis, deforestation, global disease, economic instability, tourism, migration and so on. Yet alongside these matters of international importance, it might equally be shown that geography is concerned with the everyday, the local and (sometimes) the seemingly banal. After all, geography has always been concerned with documenting the things that people do on a day-to-day basis, noting variations across space in the way that people work, rest and play. As such, activities such as eating, shopping, conversing, walking, playing, sleeping and recreation are all inherently geographical, in the sense that they can be understood as practices on which the discipline of geography can offer a distinctive perspective. To show this, we can take a seemingly mundane news story plucked from the relative obscurity of p. 7 of the *Guardian* of 29 September 1999 (see Figure 1.1). Written by Lucy Ward and Keith Harper, this describes Labour politician John Prescott's plans to designate new National Parks in the UK. Read the story, and think about how it begins to open up some different geographical dimensions.

Perhaps the first thing we might note here is that this story does not concern a matter of pressing global importance (although it might well be of crucial importance for some). Indeed, we might expect that the story would be of little or no

Figure 1.1 New National Parks in the UK . . .

Prescott to unveil new national parks

Environment

Lucy Ward and Keith Harper

John Prescott is today set to announce the creation of the first national parks in 50 years – in the South Downs and the New Forest.

The move, to be unveiled on the 50th anniversary of legislation confirming the protected status of the parks, will bring the total number of national parks in England and Wales to 13.

The establishment of a South Downs national park, in particular, will delight environmentalists, who have long called for protection for the area. But landowners say extra protection for the Downs is bureaucratic and unnecessary.

When a Labour government led by Clement Attlee brought in the 1949 national parks and access to the countryside act, the South Downs was the only proposed area to be turned down for the status.

Mr Prescott's decision to bow to pressure to bestow national park status on the Downs – straddling West and East Sussex and parts of Hampshire – will be seen as a u-turn by campaigners.

Kate Parminter, director of the Council for the Protection of Rural England, said the creation of new parks would be warmly welcomed.

She said: "If this is true, it will be a landmark decision which will do much to protect England's most beautiful and valued landscapes."

The deputy prime minister is also expected to use his speech today to demand that the 25 railway companies give a commitment to curb fare rises if they want to renew franchises from 2002. He is increasingly sensitive to public hostility to fare increases, particularly with companies such as Richard Branson's Virgin Trains, which has increased some fares by almost 40% in the past three years.

Mr Prescott wants his new strategic rail authority to insist that fare increases are held to reasonable levels.

Some of the 10 train companies which operate into London, including Connex, will have next year's fares reduced because their performance and punctuality have failed to meet government targets.

Companies within a 50 mile radius of London have their fares controlled, but no brake exists in the rest of the country.

Mr Prescott wants the industry to get rid of its fragmented tag, and will talk about "Britain's railways".

Campaigners are delighted at extra protection for the South Downs Photograph: Roger Bamber

Source: Guardian (29 September 1999); photograph by Roger Bamber

concern to anyone outside the UK. Nonetheless, we argue that this brief newspaper article serves to make the point that 'geography matters' as it is a story that is replete with geographical dimensions and ideas – many of which we will develop in this book. We begin, then, by drawing out some of the geographical themes that are inherent in this story. On the surface, the story concerns the proposed designation of two new National Parks in southern England, something commented upon firstly, by a representative of the Council for the Protection of Rural England (CPRE), and secondly, by two journalists writing in a British national newspaper. An initial thought might then be, why is this story in any sense a geographical one? Why should the planned designation of two areas of land as National Parks be of geographical concern?

Firstly, and perhaps most obviously, this is a geographical story because it concerns the protection of a landscape that has been created through physical processes. The South Downs, for instance, is a chalk downland that has been shaped and weathered through processes of erosion and deposition, that has been subject to the invasion and succession of various species of flora and fauna, and that has changed according to specific weather and climatic conditions. As such, we can understand the story as one concerning the environmental processes that culminate in the creation of these distinctive physical areas. But while this is one take on the story, it is a partial one in the sense that these landscapes are also **peopled** landscapes – places that have been created, transformed and used by people in a variety of ways over time. As human geographers we can thus develop different angles on this story, some of which rely on our understanding of the national context in which the article has emerged, and require us as geographers to think about its wider resonances as well as its more explicit content.

We could start, for instance, by saying something about how the apparent need for areas of land to be offered 'protection' is associated with particular forms of human habit or **behaviour**. In the UK, many people like to spend leisure time in areas which are often geographically separated from places associated with other aspects of their lives (such as places of work or education), often exacerbating environmental pressures on areas regarded as being of natural beauty. We might, therefore, begin to think about the ways people organize their use of space and time, making decisions about when and where they will spend their leisure time. From a geographer's point of view, we might be interested as to why areas like the South Downs or the New Forest are subject to particular pressures from mass recreation, provoking the application of legislative protection to preserve the character of these 'special' areas. Here, we can also take a temporal perspective to speculate on why these types of landscape are deemed as being worthy of protection now (rather than 50 or even 150 years ago). Is this because an increasing number of people now perceive these areas as offering better opportunities for recreation and leisure than they did in the past? Or is it because these landscapes have taken on an enhanced value relative to the type of landscapes that have been considered worthy of protection in the past? These are not easy questions to answer (see McEwen and McEwen, 1987; Winter, 1996), but certainly geographers have sought to collect data to examine both viewpoints. For instance, noting changes in the way these areas have been painted and written about might allow us to examine how specific landscapes became highly valued (see Chapters 6 and 7), while interviews and surveys with users of these places might allow us to develop an insight into what people think of these places (Chapter 3). The ways in which people **perceive** or **imagine** particular types of place (as places of beauty, leisure, rest or play) become central to our understanding of human geography, helping us to understand why some areas become the focus of particular types of human activity.

A further geographical dimension we can draw from this story is that people seem to become attached to specific places. Places take on a significance far greater than their simply being locations on the Earth's surface. It is clear from the article that for some people, the South Downs is a very special place. Consequently, another geographical response to this story might be to suggest that these places are **meaningful** to people, being multidimensional in the range of meanings and significancies they can carry (see Chapter 4). But something else to notice is that the designation of the National Park is not universally regarded as a 'good thing'. Some groups, such as the landowners mentioned, regard the designation as unnecessary, perhaps fearing that new restrictions would place limits on their own use of space and encourage an increase in the numbers of visitors. In this story, places are therefore something which can be fought over, necessarily bound up with the **power relations** that exist between different groups (see Chapters 8 and 9). In this case, the South Downs is symbolically important within the wider debates which seek to define places as either public (to be enjoyed by the masses during their recreation) or private (upholding the landowners' right to manage and use their own land). While walkers and cyclists may want to fight for increased access to the countryside, landowners may assert their right to exclude people from 'their' private property.

This raises yet another perspective – that place can be understood in relation to group concerns as well as individual desires and needs. Indeed, examining the proposed South Downs National Park as a site of potential conflict requires us to identify the involvement of a range of **interest groups**. Specifically, the CPRE is mentioned here as an organization concerned with protecting the countryside. This organization was formed during the late nineteenth century as part of a reaction to expansion of urban sprawl into the British countryside, campaigning for the protection of rural places along with what was imagined as a particularly rural social order. This was, supposedly, a good, natural and harmonious social order based on an appropriate degree of deference to one's 'betters', where one 'knew one's place'. We can, perhaps, take this a step further and suggest that organizations like the CPRE have been concerned with the maintenance of **boundaries**. Boundaries in themselves are important concepts in human geography (and much will be said about them in later chapters). Indeed, the construction of boundaries is central to the concept of a National Park, an area 'marked off' as worthy of protection. Here, however, the lines are not simply physical divides such as fences, walls and barbed wire (though these are important). Instead, **imagined** boundaries come to the fore as central concepts in our understanding of the relationships between people and places. As David Matless (1990) suggests, the CPRE (and agencies like it) do not simply want to protect the countryside, but seek rather to maintain an imagined distinction between the countryside and the town. Similarly, the categories of 'town' and 'countryside' need to be regarded as products of a particular geographical vision which, in itself, desires order and segregation (see Chapter 6). The fact that this sense of order can be disrupted every time an urban dweller drives out to the countryside to 'consume' the rural, or rural dwellers drive into town, watch a film, buy wellington boots or sell cattle at the market, suggests that it is fundamentally important (and easily disturbed – see Chapter 9).

There are yet other important geographical dimensions spilling out of this brief newspaper article. The demand for National Parks in the UK, although having roots in the nineteenth century, took on particular importance during the Second World War (1939–45) as part of a morale-boosting programme designed to encourage British troops and civilians by providing a vision of what they were fighting for (Evans, 1997). As the press cutting (Figure 1.1) says, legislation for the protection of the countryside was passed by the government soon after the war, in 1949. We might be able to suggest, then, that the UK's National Parks are tied to an idea of **national identity**. National Parks, the countryside they claim to protect, and the independence of a 'proud island race' that so appeals to some groups in British society, are all bound up with an idea of a national identity that supposedly unifies the very disparate groups of people living within its boundaries (see Chapter 5). Of course, there will always be competing voices whose idea of national identity differs from this very English vision. These groups may well find themselves excluded from Anglo-White versions of what it means to be British, something which may be evidenced when they find themselves in places associated with 'British values'. Indeed, some geographers have noted that in the British countryside, black and Asian people may be made to feel 'out of place' in a space that has become strongly associated with white European people (Kinsman, 1995; see Chapter 7).

The idea that perceptions exist concerning appropriate and inappropriate sorts of people in different places (and, by extension, appropriate and inappropriate sorts of behaviour), leads us further to argue that there is also a **moral** dimension to human geographies. The implication here is that certain people, and particular activities, can be considered as 'in place' or 'out of place' in specific areas (Cresswell, 1996). A geographical perspective on this might be to examine how such definitions of rightness and wrongness come into existence, for instance, by exploring the way that National Parks have been imagined as suitable for certain groups and behaviours (but not others). David Matless's (1995, 1998) fascinating account of the portrayal (on map covers and in photographs, for example) of outdoor leisure in the early twentieth century points to how particular ways of being and moving in the countryside have come to be considered as particularly healthy, as morally and spiritually uplifting, tied into particular ideas about what it means to be a 'good citizen'. Alternative behaviours, from dropping litter to becoming a 'New Age Traveller', become inappropriate as far as majority opinion is concerned. A question we might pose here is 'why?'; why is leaving crisp packets and camping in an old bus in the British countryside constructed as 'wrong'? And why is walking in the open air regarded as morally and spiritually uplifting?

Again, there can be no easy answers, and although we will suggest ways of thinking through these kinds of questions throughout this book, we need to acknowledge that human geographers must confront a whole series of difficult questions as they seek to make sense of the relations between people and place. Indeed, another dimension we want to pull out of the National Parks newspaper cutting is the question of what we might mean by 'place' anyway. The designation of new National Parks in the UK might lead us to conceptualize places as things which can be seen as independent, as bounded, and as separate from other places. A National Park, existing inside a line drawn on a map (and indicated as one drives into it by road signs) might give us the impression of a place separate from all others. What we have suggested above already contradicts this; we have shown, for example, how the use and occupation of National Parks is related to wider processes of urbanization and the impacts of town and country planning. But places are also open to influences on a far greater scale than the national. The concept of the national park itself originated outside of the UK, in the USA, where the first parks (e.g. Yosemite and Yellowstone) were set up in the 1860s and 1870s in response to concerns over the loss of wilderness (Pepper, 1996). Places, then, can be thought about as open; that is, as receptive to ideas, people and power relations extending way beyond them (Massey, 1995a). In an era characterized by

globalization, we need to be able to think about what happens to individual and distinctive places when there are powerful economic, political and social processes which, some have argued, are involved in the homogenization of place – everywhere might seem to be becoming the same (see Chapter 2). While taking issue with this claim, we recognize the importance of thinking globally, while contemplating the specificity of individual places.

So, even a relatively innocuous and unimportant newspaper cutting can be seen to be full of geographical connotations and possible interpretations. A final take we might offer here comes from the fact that the story is culled from the pages of a daily national newspaper (the *Guardian*). This is a form of **media** that is produced in certain places (head offices, printing works) to be consumed by its readers in other places (in the home, on the street, in a café, etc). What is important here is that such media do a fundamental job in circulating images and stories about places (some of which we will never experience first-hand). While many journalists pride themselves on the quality of their journalism, it is impossible to imagine that such representations can ever be devoid of bias. No doubt it would be easy for journalists to represent the new National Parks in a very positive light (or, for that matter, in a more negative light), depending on the audience for which they were writing. The media, by picking up on particular stories and telling them in particular ways, thus define the news agenda, producing the 'facts' of the stories, that, in fact, represent partial (and often politicized) ways of thinking. Thinking about the ways in which people and places are **represented**, therefore, offers yet another way of understanding human geographies and the relationships between people and place (see Chapter 7).

■ 1.2 Thinking geographically

Picking up on a relatively mundane newspaper story such as the one above has allowed us to indicate some of the key directions this book will take in its exploration of people and place, and the relationships between the two. Walking, driving or cycling in the countryside is something that many of us have done, perhaps on innumerable occasions, yet it is probably not something that we have reflected upon as being inherently geographical. Yet the story of the National Park considered above has suggested that even ordinary, everyday acts like this can be understood in a number of ways in terms of the geographical relationships between people and place. As we have seen, this is not merely because the story concerns human–environment interrelations (although this represents one way of thinking about human geography) but also because it concerns the way place is made meaningful by different groups, imbued with moral associations, used and occupied by different actors, represented in the media and so on.

This brings us to one of the principal intentions of the book: to show that behind our everyday 'being-in-the-world' (what we do, what we experience, how we physically interact with the world and its people) are extraordinary sets of relationships between people and places which have become the foci for much contemporary geographical enquiry. Connected to this (and implicit in the discussion of the National Park story) is the fact that we can understand the terms 'people' and 'place' in a number of disparate ways. Both these terms have been used very differently by geographers throughout the discipline's history. Conflicting understandings of 'people' and 'place' will therefore be drawn out during the course of the book, demonstrating that both terms are far from straightforward. People, for example, can be thought of as objects behaving in particular ways (taking part in leisure and work activities), as being able to experience attachment to place, as having aesthetic appreciation, as beings concurring with particular notions of morality or as playing roles in political and social struggle. Place is also a term that defies easy definition, being depicted variously as a 'bounded' location, as a space of flows (i.e. open to variable external social, economic and political influences), as a locale defined through people's subjective feelings, as the context for social and political relations or as a place created through media images. Of course, some of these notions are widely recognized beyond the discipline, and we might also think here about the way that geographical concepts like place are central to the language that we use to communicate with each other (for instance, when we talk about a particular sort of behaviour, activity or development being 'in place' or 'out of place' in a National Park).

Throughout this book, we will therefore begin to see that human geographers have used the terms 'people' and 'place' in a bewildering variety of ways. Although you may prefer one or other of the definitions, a central argument here is that none of them is necessarily wrong. As we will see, each represents a rather different way of approaching human geography, presenting a specific idea about the type of relationship that exists between people and place. Moreover, fundamental to human geography is the idea that we cannot study people and places independently of each other. Indeed, it is the **relationality** of people and places that is so important to geographical understanding. Thus, when we talk about 'people and place', the **and** is as important and ambiguous as the definition of the people or the place concerned. As a discipline which focuses on the role of space and place, it has become axiomatic within geography that as people construct places, places construct people (inferring a reciprocity between people and place). It is difficult, for example, to think about a place which is designated as a National Park without thinking about how that designation is reliant on the involvement of people (e.g. policy-makers, walkers, poets, etc.). At the same time, the National Park is involved with the construction of particular human identities and activities – it gives the politician somewhere to legislate about, the walker somewhere to walk and experience solitude, scenic beauty and physical exertion, and the poet somewhere to lyricize. Consequently, people and places derive their identities from each other to a significant extent. Going a step further, we also need to emphasize that this relationality is not something which is fixed and unchanging. Relationships between people and places are always in a state of

becoming rather than of simply 'being'. To adopt Marcus Doel's (1999) turn of phrase, things become what they will have been – there is no point at which we can just stop things and unproblematically capture a snapshot of people–place relationships. The relationships, and the people and places themselves, will have moved on before we know it.

We do not wish to pretend that exploring the relationships between people and place is straightforward – although we do want to suggest that it can be fascinating and illuminating, as innumerable geographers have discovered. Accordingly, subsequent chapters of the book will examine the cases for understanding people, places and the relationships between them in particular ways. Each approach to human geography represents one possible way of thinking about people and place, based on certain ideas about what it truly means to be human. Our overview cannot be anything other than partial, of course, and there are other important approaches to human geography that we will not consider here. Instead, we will foreground a number of different approaches which have focused primarily on the way that **individuals** interact with their surroundings in their everyday lives, highlighting the work of those human geographers who have sought to develop distinctive and innovative ways of studying the relationships between people and place. We suggest that each of these ideas/authors has some insight into this relationship, though they may be imagining the nature of the relationship very differently from other geographers. Inevitably, there is often conflict evident between these different ideas – with as many approaches to human geography existing as there are human geographers – but our intention is not to argue that one approach is better or more important than another (and at this point, if you haven't yet done so, you might want to go back and read the Preface, which will tell you some important things about the rationale and scope of this book, and about some things you need to know when reading this book). Our emphasis is instead that this book should be **used**; that the ideas we will explore in this book are just waiting to be lifted off the pages and 'tried out' in your everyday lives. As you read the book, therefore, you should think about how the different ways of conceptualizing the relationships between people and place help you think through our everyday lives. Do any of these ideas help you make sense of how you live your life? Which human geographers' ideas do you think are relevant to your own experiences, aspirations, needs and wants? What should a truly 'human' geography be like? These are all big questions – ones that you will hopefully develop your own responses to – but before we start to think through them, we should perhaps set the scene by rehearsing something of the development of human geography.

1.3 Approaches to human geography

Trying to document the history of human geography is something of an impossible task, and to do so inevitably involves making a number of generalizations and skimming over certain complexities in the geographical literature (Livingstone, 1992). Yet it is frequently the case that geographers have sought to tell the story of the discipline's development making references to discrete episodes or eras in the pursuit of geographical knowledge. Following the philosopher Kuhn, these are often termed **paradigms** – periods where the exploration of human geography has been driven by an overarching idea of what is the best way of 'doing' human geography. Periodic shifts between paradigms reflect attempts by geographers constantly to refine and redefine their discipline in order to demonstrate its intellectual worth. For example, in his wonderfully wide-ranging book *Geographical Imaginations*, Derek Gregory (1994) presents a schematic plan illustrating the way in which Anglo-American academic geographers engaged with ideas from anthropology, sociology and economics in distinct periods. According to Gregory, this three-stage progression involved geographers moving from (an anthropologically inspired) study of exotic cultures in the 'era of exploration' to a concern with the urban sociology of industrial cities in the nineteenth and early twentieth centuries, before developing forms of spatial analysis inspired by economics in the mid twentieth century. From this point, he argues that geography has revisited these three phases in reverse order, developing Marxist political-economy approaches in the 1970s, critical social theories in the 1980s and returning to engage with critical cultural theory in the 1990s (see Gregory, 1994).

Although useful as an heuristic (that is, as something to stimulate thought or discussion), this model is perhaps rather simplistic in its implications that first, all geographers have followed the same route through time, and second, that the approaches are unproblematically distinct from each other. In fact, it is possible to identify elements of all six of Gregory's strands in contemporary geography, and many examples where they creatively overlap. Nonetheless, we want briefly to dwell on models such as Gregory's, as they do suggest that at particular times geography has been dominated by a particular notion of what is the best way of doing human geography. While others would no doubt disagree with the fine points of Gregory's classification (see Johnston, 1991; Peet, 1998), there is at least substantial agreement that in the 1950s and 1960s, the methodology and substance of human geography began to change significantly as various geographers began to argue for a systematic human geography based on principles of experimentation and quantification. In latter years this has been described as an attempt by geographers to reorientate the discipline as a **spatial science**. One of the prominent features of this endeavour was the attempt to conceive of geography as concerned with the formulation of laws explaining the distribution of phenomena on the Earth's surface.

While its origins are complex, spatial science is often seen to have emerged in the 1950s and 1960s as a reaction to preceding forms of descriptive, regional geography which tended to assume that the way people lived was largely a function of the distinctive physical environments in which they found themselves. As Ellen Semple (1911, 1) put it, 'man [*sic*] is a product of the Earth's surface'. In its most extreme form, such regional geography was associated with 'environmental determinism' (that is, the environment *determined* how people lived), and could be used to justify racist attitudes (such as the idea that African people were lazy and lacking intelligence 'because' of the environmental conditions in which they lived) (see Duncan, 1993, for a more detailed discussion). At the same time that the problems with this type of approach were being recognized, there was pressure on geography (and other social sciences) to become more like a 'proper' and 'useful' science. This implied a mode of geographical enquiry determined to search for rules and laws, making use of numbers and statistics. The term '**quantitative revolution**' has subsequently been applied to describe what happened to much academic geography in this period (to 'quantify' means to be able to measure phenomena or events in numerical terms). Geographers, in order to make their subject respectable in a context where scientific explanation was increasingly dominant, created a 'notion of geography as a spatial science concerned with modelling and predicting human spatial behaviour' (Cloke *et al.*, 1991, 13). Thus, where 'natural' sciences like physics and chemistry tried to create general laws and rules about things like molecular structures and chemical reactions, geographers were interested in creating models of spatial structures which could, for example, generalize settlement patterns, urban structures or industrial location. Simultaneously, many human geographers became concerned to adopt the 'scientific methods' of, for example, hypothesis-testing. What was more, the laws that geographers produced often adopted scientific language (e.g. using 'gravity models' and 'spatial physics' to predict flows of commuters between settlements of different sizes) in order to underpin this way of thinking about human geography.

In this story of geography's 'scientific turn', the movement towards quantification and spatial analysis triggered a subsequent critique based around the fact that the notion of geography as constituting a search for spatial laws resulted in abstract models and ideas that bore little relation to the complexity of the 'real' world. Recognizing that it was in fact very rare that settlement patterns or traffic flows ever did conform with the models and equations of spatial science, they often implied that if things were properly organized then that was what the world *ought* to be like! That is, such models became **normative** prescriptions, perhaps useful to policymakers and planners in their attempts to order and impose discipline on the chaotic world around them. During this period, then, much geography was an obsession with collecting numbers, counting things, and performing complicated statistical operations to try to model what was going on (or ought to be going on) on the Earth's surface.

In relation to the ideas we address in this book, a crucial question concerns how this approach conceptualized people and place. Going back to our starting topic, National Parks and countryside recreation, consider the following gravity model (Coppock and Duffield, 1975) which attempts to predict the number of trips made by people to a recreation site from a settlement (origin). This is the one and only (statistical) equation which appears in this book! Think about the ways in which places and people are represented in such a model.

$T_{ij} = KO_iA_jF(C_{ij})$
where:

T_{ij}	= trips between origin i and recreation site j
O_i	= number of trips generated in origin i
A_j	= the attractiveness of recreation area j
$F(C_{ij})$	= a function that expresses the 'friction' involved in overcoming the distance between O_i and A_j
K	= constant

(Coppock and Duffield, 1975)

Starting with the concept of place, if a spatial scientist is trying to work out general laws of human spatial behaviour, then the specificity of individual places (those things that make a place special or unique) are not important. Instead, places are seen merely as background 'noise' or 'interference' which disrupts the smooth running of the normative model. Based on an assumed 'isotropic' surface (a featureless plain), places are represented as letters (i and j) whose characteristics, such as 'attractiveness', can simply be reduced to a number (A). This represents an important criticism of spatial science, in that it could be accused of taking deeper understandings of place out of geography. In the abstract and highly stylized models developed by spatial scientists, places are, in effect, effaced, replaced by a geometrical matrix of movements, channels, hierarchies, nodes and surfaces (Haggett, 1990).

The proponents of spatial science argued that an understanding of spatial structures through descriptive mathematics and modelling would lead to an enhanced understanding of spatial organization and human activity. Yet criticisms of this approach became increasingly widespread in the 1960s as the assumptions of spatial science were bought into question. One of the most obvious critiques was that the isotropic landscapes assumed by spatial science simply did not exist. More fundamental, perhaps, was the emerging criticism that spatial science worked with a very limited view of what it is to be human. In many of the models developed by spatial scientists, people were frequently represented as vectors or movements (making up aggregated flows). In the model given above, for example, people are regarded as a flow of 'trips' (T). In other models, people were conceived in very simple terms as 'economic men' who always acted in an entirely rational manner in order to reduce their expenditure of effort and maximize their economic well-being. One important consequence of this approach was that geography was very much inspired by particular packages of economic theory (and here we are in the third stage of Gregory's schematic development of geography). These tended to deal with that which is quantifiable, such as units of currency, profits, losses, amounts of resources and numbers of labourers. In such an economic frame of mind, people were seen merely as economic units or parts of a production process, rather than as complex beings who can experience love, happiness, beauty, pain, injustice or oppression. In retrospect then, this was a new human geography that 'borrowed' some interesting concepts from other disciplines, but in so doing filtered out some other interesting ideas that offered different perspectives on people and place. It is these ideas that form the basis of this book, as we will outline below.

1.4 Geographies of people and place?

In the 'history' of different approaches to human geography, spatial science is taken to represent an important and distinctive phase of the discipline's development, albeit one that is often portrayed as encouraging an arid (and ultimately pointless) human geography. Here, we would discourage that type of interpretation, and instead choose to highlight the contribution that spatial scientists made (and indeed still make) to our understanding of spatial decision-making. But our brief account of the features of this approach should begin to explain why some geographers reacted against the turn to spatial analysis. By developing analytical models based on economically- and spatially-rational human behaviour, many of these geographers populated their books and articles with faceless, characterless, emotionless units, moving round rationally on a featureless isotropic surface. This was 'human' geography only to the extent that people were considered on an aggregate level as rational actors rather than as active, emotional or creative human beings. In effect, people were 'aggregated out of existence' in this approach, while 'places' were effaced in favour of geometric nodes.

In sum, the spatial science approach clearly offered (and still offers) geographers a way of thinking about human spatial behaviour and areal differentiation based on harnessing the analytical potential of quantification and mathematical modelling. But the history of geography suggests that this approach has been rejected by subsequent generations of geographers who, for better or worse, have sought to develop more 'human' alternatives. The rest of this book is therefore dedicated to exploring how people have been (re)inserted into human geography by many of these writers. Although we cannot give more than lip-service to many of the ideas that have been developed by human geographers in the last 30 years, our primary task here is to illustrate some of the ways that geographers have reacted against the dehumanizing elements of spatial science, looking at how they have developed alternative models of humanity in the search for different (and perhaps better) understandings of the relations between people and place. As Gregory (1994) suggests, this search has resulted in renewed engagements with sociology and cultural studies, not to mention psychology, anthropology and other human sciences, as we will see. At the same time, many of these attempts to re-people human geography have gone hand-in-hand with a reconsideration of what 'places' are (e.g. things which are meaningful and multidimensional rather than simply points on an isotropic surface).

This book is thus divided into a number of chapters which, while designed to stand alone, can also be read in sequence as a series of different approaches to studying the relations between people and place. As we highlighted in the Preface, each chapter attempts to summarize one particular way of thinking about the relationships which are played out between people and place on an everyday basis. While there are many overlaps, discontinuities and gaps between them, the following provides a rough indication of the content of each chapter:

- Chapter 2 begins by exploring the concept of **globalization** suggesting that it is rapidly changing the nature of relations between people and place. Nevertheless, people's lives are conducted in **localized** sets of places, and the chapter explores ideas of space–time routines, demonstrating how everyday activities are centred on a range of places which are so close, so familiar and so fundamental to our lives that their very importance is often forgotten.
- Chapter 3 extends this discussion of everyday activities in specific places by looking at what have come to be called **behavioural** contributions to human geography. Behavioural geography highlights the complex way in which we look at, make sense of and remember the arrangement of places across space, drawing heavily on psychological research which has explored how we perceive and understand our surroundings. We stress that there are individual differences in spatial knowledge and understanding, taking examples from different social groups and different behavioural environments.
- Chapter 4 focuses on the way in which people form emotional attachments to places. Such **humanistic** approaches have examined the way in which places become meaningful to people as they encounter them in their lives. These ideas are extended by looking at the concepts of 'home' and 'landscape' in humanistic geography.
- Chapter 5 starts to question some of the assumptions made in Chapter 4, as, for example, it is shown that 'home' is not always a wholly positive concept. We try to show why some places evoke strong emotions of fear in particular groups, and raise questions about the nature of territoriality and national identity.
- Chapter 6 extends ideas about the meanings people attach to different places. We consider how places are **imagined**, for example as centres or peripheries – places which symbolize power and distinction and those which are on the imagined margins of society and space. Examples of place imagery are highlighted here, with myths of the city, wilderness and 'the Orient' explored to investigate how social groups imagine their surroundings.
- Chapter 7 focuses on the complex way that people **represent** place through a variety of media such as poetry, music, art, film and television. We show how places such as the countryside are inevitably represented selectively and that 'common-sense' views of place are bound into ideological structures. Finally, the role that mapping plays in representation of place is considered, emphasizing that map-makers play a key role in representing the world in a selective manner.

- Chapter 8 analyses the way in which places are shaped by flows and structures of **power**, by demonstrating how individuals are controlled in places in ways which promote certain cultural values and lifestyles. Examples at a variety of spatial scales, from the home to the shopping centre, from the classroom to the city, are examined to explore the contention that people's behaviour in place is subject to sometimes subtle forms of discipline.
- Chapter 9 argues that places (and the meaning of places) are subject to **contestation** and **negotiation** between different cultures and lifestyles. We investigate the cultural politics of place, with examples from urban development and the occupation of the streets, for example, used to demonstrate the ways that multiple senses of place are worked through in complex and sometimes contradictory ways.
- Finally, Chapter 10 avoids simply summarizing the content of the previous chapters, and instead attempts to question the **philosophical**, **pragmatic** and **political** underpinnings of the various approaches adopted in the examination of relationships between people and place. We argue that each approach is based on a set of assumptions which need to be carefully considered prior to undertaking geographical research. We try to avoid vague assertions that certain approaches are 'better' in particular situations, and suggest that different aspects of the nature of everyday life can be illuminated through the adoption of different perspectives. We hope that this chapter highlights a series of possible points of departure for your own geographical studies.

■ Summary

In this chapter we have begun to explain why a geographical take on what occurs around us is useful – and perhaps essential – for making sense of human relationships. In particular, we have highlighted the following key ideas:

- Human geography necessarily involves the investigation of relationship between people and place, with different ideas of what 'people' are and what 'place' is having been adopted by different geographers at different times.
- For some geographers, a useful way of understanding the relationship between people and place has been to reduce human beings to abstract points on a surface whose behaviour can be mapped, modelled and predicted.
- In contrast with (and reaction to) this notion of geography as spatial science, other geographers have sought to develop different ways of thinking about people and place, creating a variety of (often conflicting) 'human-centred' geographies.

two Everyday places, ordinary lives

This chapter covers:

2.1 Introduction

2.2 Space, time and globalization

2.3 Mapping the geographies of everyday life

2.1 Introduction

In the last chapter we discussed the idea that human geography may be defined, amongst other things, as involving an examination of the type of relationships that exist between people and places. All too often, however, any sensitivity to the complexity of the relationships which people have with their surroundings has been lost by geographers effectively taking the people out of human geography. As we discussed, this tendency to **dehumanize** human geography was perhaps most pronounced in the 1950s and 1960s as certain groups of geographers sought to construct a discipline that was capable of mapping and predicting human behaviour at an aggregate level, developing models of migration, industrial location, agricultural activity and settlement location that denigrated the role of individuals and largely ignored the capabilities of people to act as agents and initiators of change. On the other hand, we suggested that a truly 'human' geography might be imagined as one which is aware of the capabilities and potential of human creativity, is mindful of how individuals make sense of their surroundings and exhibits a sensitivity to human difference and diversity.

In this chapter, therefore, we want to begin to explore what a 'human-centred' human geography might consist of by exploring the different **scales** at which we might make sense of human agents and their relationships with place. Here, we are using the term scale not so much in its cartographic sense (where the distance on a map corresponds to a fixed distance in the real world), but in the sense that geographers have sought to investigate the relations between people and place at a variety of distinct but hierarchically ordered levels (such as the global, the transnational, the national, the regional and the local). As the nature of the relationships occurring between people and place at one scale are always influenced by the nature of those occurring at other scales, we want to start this chapter by examining how various social, economic, political and cultural trends are encouraging geographers to explore these relationships at the **global** level. As we shall see, an awareness of global processes is necessary if we are to understand the complex nature of relations between people and place. However, we want to argue that this is not in itself sufficient for explaining the fundamentally important role of **place** in shaping people's routine, 'ordinary' lives. Instead, we argue that such an exploration needs to be grounded in an investigation into the way that people's movements and behaviours centre on a set of local, 'everyday' places whose importance has often been ignored or downplayed in geographers' rush to develop large-scale 'grand' theories. In developing this argument, the latter half of this chapter turns to examine what we might mean when we refer to the geographies of everyday life, and to establish how a focus on the extraordinariness of ordinary life offers a unique insight into the relations between people and place. To begin with, however, we return to the basic concepts of 'space' and 'time' in order to question what might be meant by these terms in an era characterized by dramatic global change.

2.2 Space, time and globalization

In Chapter 1, we hope we gave the impression that geography is a discipline which has been characterized by intellectual trends and fashions that impel geographers to write about particular things in particular ways at particular times. While it is dangerous to overemphasize the importance of these fashions – not all geographers will be swept along in the general enthusiasm for a new-found perspective or intellectual fad – it is clear that there has been widespread interest over the last decade in thinking and writing about global issues. While this focus on the global is far from new (with the foundations of geography often having been traced back to those cartographers and explorers who sought to identify the distinctive environments, climates and cultures associated with specific parts of the world – see Livingstone, 1992), a predominant trend in current scholarship is to explore the causes and consequences of **globalization.** This is a term whose precise meaning remains the subject of considerable debate, but it is commonly used to refer to the process (or processes) whereby all parts of the world are becoming subject to the same sorts of influences. Here, economic, social and political institutions, flows and networks are all regarded as significant in linking the world together in single 'world-system', creating a global situation where what occurs in one nation inevitably impacts on other nations (with the nation-state consequently seen as of diminishing importance as a political force).

On one level, this globalization thesis appears to present a profound challenge to 'traditional' geographical ideas that the 'spacing' and distance between different locations influences the level of interaction and exchange that occurs between them. Indeed, the basis of much geographical understanding is that flows of products, information and people diminish as distance between two locations increases. The idea that physical distance acts as a limiting effect on interactions of all sorts is captured in the concept of **friction of distance** (i.e. the increased resistance an object encounters when moving longer distances). Yet over time, friction of distance has been effectively reduced through improvements in transport and technology. This has meant that distance has become less of an obstacle to movement and communication, with the distances between different places declining in significance. Donald Janelle (1969) referred to this process as **time–space convergence**, a term which captures the apparent 'shrinking' of the world occurring as the time taken to travel through space diminishes. This shrinkage has been facilitated through a combination of technological innovations and transport improvements which, according to Harvey (1989), have occurred in two main phases. The first was at the end of the nineteenth century, as new 'enabling' technologies such as the telegraph, telephone and rail and steam ships permitted speedier flows of goods and faster communications. While this led to an immediate rise in international trade, it has arguably been in the last 30 years that developments in commercial air travel, satellite and microwave technology, fibre-optics and computing have effected a global communications revolution. At the heart of this revolution is the development of the Internet and the World Wide Web, an 'Infobahn' that allows virtually instantaneous transmission of text, pictures and sounds across the world (see Figure 2.1). At the same time, many people now possess mobile telephones which allow them to contact people around the world at a fraction of the cost that their parents' generation would have expected to have paid.

As such, the connections between different localities and nations have intensified dramatically over recent years, leaving a situation where we are seeing unquestionably faster flows of **goods**, **information** and **people** around the world than ever before. In the 1960s the cultural theorist Marshall McLuhen spoke prophetically of the coming of a 'global village' (McLuhen, 1964); by the 1990s it seemed that this global village had arrived. Today, we live in a world where momentous world events can be beamed into our homes via satellite technology; where products and ideas from abroad are routinely incorporated into our lives; where the political decisions that shape our destinies are as likely to be debated overseas as in our own governments; and where many people expect to work outside their country of birth for long periods. In short, it seems that we are living in a world where interactions of all types have been 'stretched' over space to the extent that formerly autonomous nations and locales have been drawn into contact with one another.

Here, it needs to be noted that this increased global interaction has in part been influenced by shifts in the nature of international economic activity. Hence, while commodities and products have been traded *between* nations for hundreds and perhaps thousands of years, today many of these international transactions occur *within one company*. Indeed, perhaps the decisive factor underpinning globalization is that a major increase has occurred in the proportion of world trade that is conducted by transnational corporations (or TNCs). A TNC is any company that has

Figure 2.1 Growth of the Internet and World Wide Web

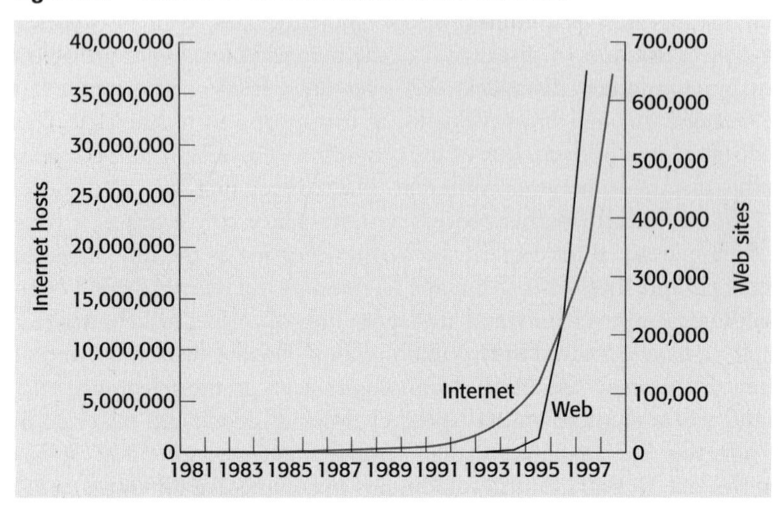

Source: Short and Kim (1999)

investments and activities that span international boundaries and that has production facilities (factories, assembly plants, offices or research establishments) in several nations. Many of the largest TNCs (see Table 2.1) are household names throughout the world because they produce so-called **global products** – a phrase used to refer to those consumer and household goods which are produced, marketed and sold under the same branding world-wide. Notable examples here include Coca Cola, Benetton clothing, Marlboro cigarettes, Nike trainers, the Sony Walkman, the Microsoft Windows operating system and Ford motor cars. Indeed, a widely noted trend is for these products to be marketed using images and discourses of globalization, with consumers sold the idea that buying a particular product connects them into a **global culture** (Myers, 1998). This is often defined by advertisers around Western ideals of environmental concern, respect for cultural difference and (perhaps paradoxically) the celebration of conspicuous consumption (see Figure 2.2).

Beyond these manufacturers of mass-produced consumer goods, some of the most powerful TNCs are involved in oil and petrochemical production, pharmaceuticals and aerospace. Yet it is also important to note that many TNCs are service sector-based (with services constituting the fastest growing sector of overseas investment). Accounting, advertising, financial and legal services are of major importance here, with these **producer services** playing a crucial role in maintaining a globally integrated financial system which sees billions of dollars of stocks, shares, bonds and futures traded every day. Processes of financial homogenization in the late twentieth century are conspiring to make national financial systems increasingly similar, as witnessed in the equalization of short-term interest rates and the introduction of standardized forms of market regulation (Leyshon and Thrift, 1994), and,

Table 2.1 The ten biggest TNCs by market value (US$), not adjusted for inflation

Company (country of origin)	Market value (billions)
Microsoft (United States)	$406.22
General Electric (United States)	$333.05
IBM (United States)	$214.81
Exxon (United States)	$193.92
Royal Dutch/Shell (Netherlands/UK)	$191.32
Wal-Mart (United States)	$189.55
AT&T (United States)	$186.14
Intel (United States)	$180.24
Cisco Systems (United States)	$174.09
BP Amoco (UK)	$173.87

Source: Business Week (12 July 1999)

Figure 2.2 Global discourse as advertising strategy: the United Colors of Benetton

perhaps, in the globalization of certain currencies (especially the US dollar) and the creation of a new European currency in the European Union. A further illustration of this is evident in the advent of virtual money, which cannot be considered to belong to any particular nation. The credit cards issued by Visa, Mastercard, Barclays and American Express potentially allow all of us to benefit from this speeding up and stretching of monetary transactions. This means that it is entirely possible for us to tap into the networks of global finance every time we use an automatic cash machine, obtaining money in local currency wherever we are in the world. Thus, the need to translate paper money from one currency to another has virtually been obliterated, with the large financial corporations and institutions who manage these flows of virtual money – Nat West, Citicorp, J P Morgan, Satsura Bank, Hong Kong Shanghai Bank (HSBC) – now among the largest employers in the world.

Yet whatever business they are in, the majority of TNCs aim to have a strategic presence in the global market for a number of reasons – both **supply**- and **market**-oriented. Here, we need to distinguish between a company's urge to seek the most profitable location to make products (i.e. to seek locations where the supply of raw materials, labour and infrastructure is most conducive to producing a commodity at lowest cost) and its desire to open up profitable new markets for that product or service. An example that might help to illustrate the factors at work here is that of Nike. As the leading athletic footwear company in the world, Nike needs little introduction. In 1996 its annual revenue was reported as US$6.5 billion, a figure comparable with the gross domestic product of some small nations. Considering that the company was founded as recently as 1965 (in Beaverstown, Oregon), this represents a phenomenal expansion. In part, the company's success can be explained with reference to its international production operations which have allowed it to shift the labour-intensive processes of shoe assembly from nation to nation as local costs of production change. This has meant that virtually all of the company's shoe manufacturing is currently in the Far East, where low wages are an obvious attraction (Schoenberger, 1998). Nike's principal offshore operations are thus in five Asian countries – China, South Korea, Taiwan, Vietnam and Indonesia – where the employment of women on ultra-low wages allows the company to accelerate the rate at which it accumulates and reinvests profit. In the future, of course, it is entirely possible that this accelerated growth might be more easily achieved somewhere else, meaning that the current operations might be abandoned in favour of less expensive places. In this sense, TNCs like Nike are depicted as increasingly footloose in their search for profit, able to deploy their activities across international boundaries with seeming impunity.

Transnational corporations are active not only in manufacturing and financial services, but with improvements in communication it has been possible for media companies to extend their global reach, shaping the tastes and desires of new international audiences. While it was as recently as 1965 that the first commercial communications satellite (INTELSAT or Early Bird) was launched, today there are over

160 satellites in orbit enabling broadcasting companies to reach many nations still not served by a terrestrial television service. The world's six largest media organizations – Rupert Murdoch's News International, Time–Warner–AOL, Disney, Bertellsman, TCI and Viacom – all have a global presence, broadcasting to an international audience via their satellite subsidiaries (BSkyB Broadcasting, RTL, Fox Channel, CNN and so on). Time–Warner–AOL, for example, was created in January 2000 from a merger of the Time–Warner and AOL (America On Line) corporations to create a US$350 billion (£220 billion) business which, according to a press release, formed 'the world's first fully-integrated media and communications corporation for the Internet century' (*Guardian*, 11 January, 2000, 1), and which became the world's fourth-largest corporation. Accordingly, it is possible to discern a homogenization of global broadcasting:

> Irrespective of where they live, audiences around the world are fed a broadly similar diet of television. The same kind of programmes are scheduled at the same times of the day. . . . Soap operas and quiz shows account for most of the daytime slots while children's programmes predominate in the early evening. These are followed by family viewing, the mid-evening news, drama, sport and adult television. The significance of this standard format is that it generates demands for particular types of programming, much of which is international in origin.
>
> (Clark, 1997, 126)

Crucially, this standardized diet of images has been implicated in the production of new forms of cultural identification as people become exposed to lifestyles, activities and values which transcend the frontiers of national communities. Appadurai (1990) is thus prompted to describe the emergence of a **global mediascape** produced by flows of images and information via printed media, television, music and film. In turn, these images serve to valorize (i.e. artificially raise the value of) and promote global products, showing that increasingly dense media connections serve to entangle local and national economic and social processes in global flows of money, information and goods.

■ Reading

The complex debates around globalization have been debated in a number of contexts – Thrift (1995) and Crang (1999) are particularly recommended because of the careful attention they give to the global as a network of flows, with Shurmer-Smith and Hannam's (1994) chapter 'The world as a place' also containing some useful pointers to debates about the way in which global cultures are negotiated in the midst of these flows. More general texts on recent economic, social and political change include Johnston *et al.* (1995) and Allen and Hamnett (1995).

Much, then, has then been made of the impact of globalization, of the 'annihilation of space by time' and the increased ease of travel and communication which characterizes the contemporary world. Some commentators have even predicted the 'end of geography', foreseeing a time when all places will be similar in terms of their cultural and social character as global corporations and media empires spread similar products and images across the Earth's surface. This process of 'deterritorialization' – an 'emptying-out' of space and time – has been explained by some geographers in relation to the changing geographies of **capitalism**. As described in the writings of Karl Marx, capitalism is commonly understood as a system of economic production which thrives on the inequality and uncertainty arising from its continual need to develop production forces – labour power, raw materials, machinery, transport – to make commodities which can be sold at a profit. This system is inherently growth-oriented, being prone to crisis when the type of commodities produced can no longer be sold at a profit or when technological advance undermines existing production methods. As geographers like David Harvey and Manuel Castells have stressed, one way that capitalism can attempt to surmount problems of declining profitability is through geographic expansion, opening up new markets and shifting production processes into new territories. It was this drive for continued capital accumulation that prompted European merchants and traders to expand their trading empires into Africa, Asia and the Americas from the fifteenth century onwards, spreading the tendrils of mercantile (trading) capitalism across the globe. In this sense, the expansion of global producers in the twentieth century can be read as just the latest phase of capitalism's voracious expansion – an expansion which has involved the incorporation of most parts of the world into capitalist production and consumption practices that take shape in the West. Accordingly, it is perhaps better to refer to globalization as representing the 'Westernization' or even 'Americanization' of world culture: a process whereby all nations have seemingly been persuaded of the benefits of Western-style capitalism.

In considering this argument, we might think of the ubiquitous **global landscapes** which actively promote modern consumer practices designed to encourage and sustain global capitalism. For example, anyone who travels internationally cannot help but notice that there are many places that look similar no matter what country one is in. The French anthropologist Marc Augé (1995) offers the examples of airport departure lounges, railway terminals and shopping malls: settings which are decorated in a similar manner, have identical shops and food outlets, and are pervaded by the same piped muzak irrespective of international context. Inevitably, they will be populated by people dressed in similar ways (jeans, designer T-shirts, trainers for the young, business suits and casual wear for older people), reading internationally syndicated magazines and comics, or pausing to play video games which are the same the world over. For Augé, these settings – together with the landscapes of hotels, sports centres, restaurants and even motorways – extend capitalist consumer values into the far corners of the globe, simultaneously symbolizing the sense of speed, efficiency and supermodernity that lies at the heart of global capitalism.

Yet perhaps the most frequently cited example of a global landscape is that of the fast-food restaurant, particularly McDonald's (which has over 18,000 branches in 90 countries). With McDonald's recently expanding into the former communist bloc, some commentators have even redubbed globalization as the **McDonaldization** of society, suggesting that the fast-food restaurant has become the organizational force representing and extending the process of consumer capitalism furthest into the realm of everyday interaction and local identity: 'McDonaldization . . . is the process by which the principles of the fast-food restaurant are coming to dominate more and more sectors of American society as well as of the rest of the world' (Ritzer, 1993, 1).

Ritzer contends that everything from pizza to ice cream, from alcohol to fried chicken is now dominated by the McDonald's mentality. Seemingly, we no longer have to go to the chains; rather, they come to us, in city centres, malls, motorway service stations, schools and military bases, hospitals and airports, even aeroplanes and football stadia (indeed, McDonald's is a predominant food retailer in the UK's Millennium Dome). Moreover, Ritzer claims that McDonaldization has apparently extended its reach into areas that are increasingly remote from the heart of the fast-food business – book-retailing, fashion, child-care, sports, news-production and so on. His basis for this claim rests on the idea that McDonald's encapsulates five themes at the heart of global capitalism – efficiency, calculability, predictability, technological advancement and control.

The way that McDonald's expresses and promotes these values and ideas can be explained with reference to particular behaviours and practices associated with the chain's restaurants. For example, notions of **efficiency** are captured in the way that self-service and drive-in formats encourage people to think that the service is more streamlined, working better for them as individuals (perhaps ignoring the way that the restaurant can maintain its profits by encouraging customers to serve themselves). **Calculability** involves an emphasis on things that can be calculated, counted or quantified, representing the tendency to emphasize quantity rather than quality. Examples of this include the promotion of the 'Big Mac' and the 'Quarter-Pounder', and 99c (or 99p or 9.9F...) burger promotions, and the chain's pride (for instance) in opening its 500th restaurant in Britain. **Predictability** refers to the attempt to structure the restaurant environment to eliminate surprise and so that people know what to expect (so that the standards of cleanliness and decor in one McDonald's will be repeated next week in another). Ritzer also sees predictability as being enhanced through the replacement of (unpredictable) human labour by non-human **technology**. Everything about a Big Mac is thus prepackaged, premeasured and automatically controlled. Harnessing technology also enables a greater level of **control** to be exercised over consumers, and although they may be aware of the benefits and convenience of being able to order a Big Mac in any city in the world at any time of the day, they may be less concerned about the way that McDonald's is undermining local food cultures by spreading American diet and tastes. In the same way, therefore, it is tempting to suggest that the proliferation of other global

Figure 2.3 A Parisian street scene

Photo: Authors

landscapes represents the replacement of 'indigenous' local cultures by a seemingly rational and (super)modern global culture defined by American consumer values.

■ Exercise

Figure 2.3 shows a street scene in Paris. Examine the photo carefully. What signs are there that this is a global landscape? Among other things, you might think about the design of the buildings, signs advertising particular products, the commodities for sale in the shops, the type of clothes worn by the pedestrians and so on. Then you should start thinking about what, if anything, defines this street scene as 'French'. Are there people, commodities and images here which you would not expect to encounter in another national context?

Working through this exercise, you may have become aware of the limitations of thinking about globalization as a process which has created a homogeneous global landscape. Indeed, while most geographers accept that the possibilities now exist for commodities, people and ideas to be transported faster (and at lower cost) than ever before, few suggest that the distinctiveness of local places has been lost as a result.

Instead, geographers have been keen to stress that there is not one single global culture – nor hermetically sealed local ones – but that everywhere there is a complex interaction of local and global. In this sense, we might begin to see the street scene considered above as a composite or **hybrid**, created through the intersection of international, national and local ways of life. In the past, hybridity was something that we might have noted existing only in border areas where languages, rituals and cultures mix to create new cultural forms (exemplified by the Tex–Mex music, fashion and food that has become associated with the US–Mexican border area), but today we can arguably see hybrid cultures everywhere as global and local intertwine in sometimes unexpected ways. This notion of hybridity is one that begins to emphasize why grand theories of global change are in danger of oversimplifying the effects that globalization has on different people's lives. After all, even if McDonald's restaurants are similar everywhere, this does not mean that they are identical (e.g. McDonald's in Tokyo sells a Teriyaki McBurger); nor does it mean that McDonald's restaurants are used by the same kind of people in the same kinds of way irrespective of location. This is a point we want to explore in the rest of this chapter, where we turn to contrast geographies of globalization with those geographies which document the lived experiences of 'ordinary' people.

■ 2.3 Mapping the geographies of everyday life

On one level, the idea that space–time compression is creating a 'fast world' suggests that the relations being played out between different people are becoming increasingly **distanciated**, stretched out in time and space (Giddens, 1990). Media companies and businesses certainly give the impression that we can all tap into a global network of social interaction by buying mobile phones or logging-on to the Internet – new technologies that provide the possibility of interacting with people on the other side of the world. It is thus entirely possible to imagine that some people might have few friends that they ever meet face-to-face, but that they have a far-flung network of contacts with whom they are able to maintain strong and meaningful relationships through regular (typed or voiced) communication. For example, media stories about couples who have 'met' via the Internet and gone on to marry in 'real' space suggest that traditions of taking a boyfriend or girlfriend to a cinema, club or restaurant might be replaced by distanced courtship rituals enabled by new media.

The distinction between face-to-face interaction and interaction mediated by telephones, computers, etc. is very important. Face-to-face contacts and interactions are characterized by **synchronicity** – with participants speaking and reacting with one another simultaneously in 'real' time – while written communication and the new global communication forms such as the e-mail and bulletin board systems of the Internet are **asynchronous** (i.e. words are not heard as they are spoken, but are reproduced at some later point). Professor of Architecture William Mitchell emphasizes the freedom that the explosion of asynchronous, computer-mediated, social interaction bequeaths:

> Each morning I turn onto some nearby machine; my modest personal computer at home, a more powerful workstation in one of the offices or laboratories I frequent, or a laptop in a hotel room to log into computer mail. I click on an icon to open an 'inbox' filled with messages around the world – replies to technical questions, queries for me to answer, drafts of papers, submissions of student work, appointments, travel and meeting arrangements, bits of business, greetings, reminders, chitchat, gossip, complaints, tips, jokes, flirtation . . . If I have time before I finish gulping my coffee I also check the wire services and a couple of specialised news services to which I subscribe, then glance at the weather report.
>
> (Mitchell, 1996, 7)

Mitchell makes the point that traditionally you had to go somewhere – a square, a café, a pub, a common room, an office or a village green – to experience this type of social interaction and to wallow in this world of information and gossip. In the contemporary (global) era, however, the Internet has enabled a distinctively aspatial mode of communication to emerge which links individuals (or e-mail identities like P.Hubbard@lboro.ac.uk) at *indeterminate* locations (depending on when and where they choose to log on). Here then, distanciated communication supports the idea that a deterritorialization process is occurring, taking social life away from the 'fixities of tradition' (Giddens, 1990, 53).

Electronic media, therefore, seem to affirm McLuhen's thesis that we are witnessing the advent of the global village – a society where everyone is able to experience the meaningful and close-knit social interaction associated with 'traditional' village life but on a global scale. On the other hand, we need to acknowledge that face-to-face contact still represents the fundamental form of human interaction entered into by people as they go about their daily routines. Even the movers and shakers of the global economy – bankers and market traders – recognize the importance of such sociability, with traders seeking to develop close business contacts with other traders in their attempt to predict market trends and pick up on rumours about impending deals. While such information can be exchanged via telephone, fax and e-mail, developing such relationships is easier if people are in (physically) close proximity to one another, resulting in the tendency for established financial centres to be more (rather than less) important in the global era. As Leyshon and Thrift (1994) note, the City of London thus remains the centre of UK business in spite of the possibilities offered by electronic trading because of the desire of traders to frequent the bars, clubs and Masonic lodges where they can meet other 'high-flyers'.

The implication here is that a sensitive analysis of space and time cannot simply focus on the global, but needs to focus on the 'here and now' where local and global geographies intertwine. Accordingly, Doreen Massey suggests that we need to think about issues of geographical scale in relation to the extent that different individuals and groups tap into global flows of finance, communication and movement. As she notes, there are certainly many producers and consumers in transnational industry, modern telecommunications, and international entertainment whose lives are predominantly played out in the 'fast world'. Yet if the world is speeding-up for some, it is slowing-down for others, and there are those that Massey depicts as 'simply on the receiving end of space–time compression', referring to 'the pensioner in a bedsit in any inner city in this country eating British fish and chips from a Chinese take-away, watching US films on a Japanese TV and not daring to go out at night' (Massey, 1993, 62). Here, it is apparent that while all people's lives have been touched by globalization, they participate in and contribute to globalization processes in very different ways. Indeed, although all people 'travel' globally to some extent (not just when they go on holiday, but when they eat 'foreign' food or when they watch television), for most, mobility is relatively limited:

> Although the world is increasingly well-connected, we must hold this in balance with the observation that most people lead intensely local lives: their homes, workplaces, recreation, shopping, friends and other family are all located within a relatively small orbit. The simple and obvious point that overcoming distance requires time and money means that the everyday events of daily life are well grounded within a circumscribed arena.
>
> (Pratt and Hanson, 1994, 25)

Evidence that many lives are still focused on a relatively clustered and localized set of places has primarily been collected by geographers interested in mapping the **activity spaces** of different individuals and groups. Here, the term 'activity space' simply refers to the space in which the majority of a human being's activities are carried out. While it is impossible to define this space precisely, drawing discrete boundaries around a bounded space in which people's activities are focused, the general notion of an activity space does begin to stress that most people's daily travel is structured around a set of local movements with only occasional forays further afield (Massey, 1995a).

Some support for this view can be found in the literature on **transport geography**, wherein geographers have begun to document how access to transport may constrain (or enable) people's day-to-day behaviour in important ways. Indeed, although over one million people travel internationally every day (primarily by air), such movements are mainly limited to exceptional vacation and business trips; for most, journeys are relatively short-range, repeated and routine movements (Golledge and Stimson, 1997). Figures from the British government suggest that the 'average' adult travels six and a half thousand miles in a given year. This means that an average person moves only 18 miles per day to fulfil all their personal duties and commitments (whether to work, shopping, child-care, leisure and recreational activities) and that while the majority of this movement is car-borne, journeys by foot are not insignificant (see Table 2.2). By implication, this suggests that people tend to have a fairly restricted activity space, and do not venture far from home as a result of their day-to-day activities (although, since these are average figures, they also imply that some people are far more mobile than others). Here, it is also interesting to note related figures from the same transport survey which suggest that the activity space of the individual is dominated by movement to and from regular activity

Table 2.2 Distance travelled by the average UK person per year: National Travel Survey 1994–96

Mode of transport	Distance travelled (miles)
Walk	200
Cycle	38
Car	5,535
Public transport	738
All modes of transport (total)	6,570

Source: *Social Trends 1998*, National Statistics © Crown Copyright 2000

locations, such as journeys to work, to shop, to socialize and so on (Table 2.3). Importantly, these figures stress that activity spaces are also closely linked to an individual's role within society, with discernible differences evident between men's and women's journey types. As we will examine in Chapter 5, this gendered interpretation of activity space is important for indicating that women and men's experience and occupation of space is on very different terms (see also Laws, 1997).

Perhaps the most explicit geographic approach to exploring the shape and form of activity spaces is in the so-called **time geography** developed by geographers at Lund University, Sweden, in the 1960s. This contextual approach to understanding human spatial behaviour was based on the idea that space and time are resources on which individuals can draw in order to realize personal activities or **projects**. These projects are goal-oriented tasks which may be identified on a variety of scales (i.e. a project to get a degree may be broken up into separate projects involving

Table 2.3 Average number of journeys completed by the average person, per year: National Travel Survey 1994–96

	Males	Females	All
Commuting	18%	12%	15%
Business	5%	2%	4%
Education	6%	6%	6%
Shopping	18%	24%	21%
Other personal	20%	25%	23%
Leisure	31%	31%	31%
All	100%	100%	100%
Total journeys per year	1,084	1,033	1,057

Source: *Social Trends 1998*, National Statistics © Crown Copyright 2000

attending lectures, writing essays, reading books; shopping may be broken up into grocery-shopping, clothes-shopping, picking up clothes from the dry cleaners, etc.). Based on the assumptions that people's ability to conduct different projects at the same time is limited, that movement between different locations in space takes time and that human beings are 'elementary' and indivisible 'particles' (i.e. you can't be in two places at once), this approach sought to examine the interweaving of people and projects in coherent 'blocks' of space and time. These ideas were most fully elucidated in the work of Hägerstrand (1982), who both provided a vocabulary to describe people's use of time–space and stressed the importance of space–time in shaping the daily movements necessary for personal survival and 'development' (i.e. eating, sleeping, finding a partner, working, bringing up a child and so on). For

Figure 2.4 A time–space map of two individuals

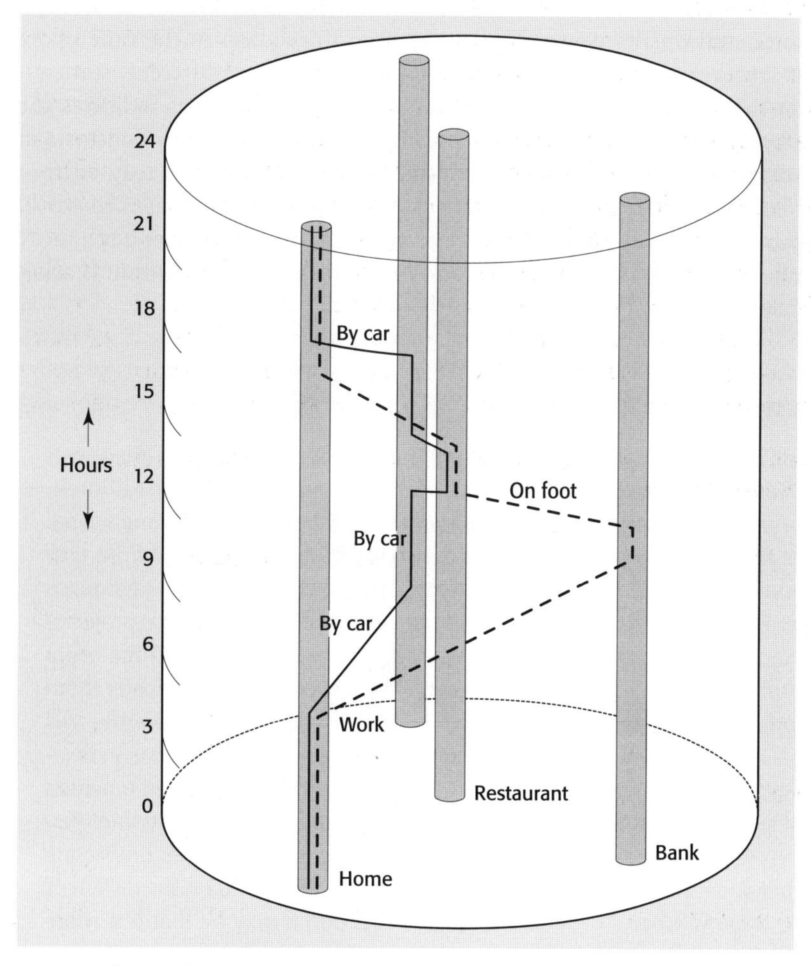

Source: Parks and Thrift (1980)

Hägerstrand it was important to demonstrate that individuals routinely and repeatedly draw upon resources of time and space in the conduct of their everyday lives.

This concern with mapping the life-paths of different individuals and groups can be illustrated with reference to one of the principal notational devices used by Hägerstrand – the **time–space map**. Such maps can be used to show the spatial extent of a person's movement during a certain time-period. This diagrammatic representation shows the scope of possible activity spaces based on the idea that a person moves between a variety of **stations** where they spend time on project-related activities. On any given day, the sum of these movements constitutes a daily path, while over longer periods daily experiences combine to produce a cumulative life-path, which simultaneously represents both the end result and context of people's actions (Pred, 1985). In Figure 2.4 each path, therefore, represents the trajectory of two individuals who live in the same house yet whose daily lives take in a number of different stations which they visit to fulfil different projects. When rendered in three dimensions, movements over space can be seen in relation to the time taken both to fulfil tasks and the time taken to move between different stations. Accordingly, in the example here, the relative 'steepness' of the lines indicates the speed or velocity of an individual's movement. Here then, we can show contrasts in the mobility of a hypothetical couple, where one partner travels primarily on foot and the other has access to a car; in the same time that one partner drives to work, eats with friends at lunchtime and returns home in the evening, the other has a more localized routine involving a visit to a bank and lunch at the local restaurant (Parkes and Thrift, 1980).

■ Exercise

Using a piece of graph paper, try to construct a space–time diagram to map your daily path. Here, it is probably easiest if you do not attempt to construct a three-dimensional map but limit yourself to a more straightforward two-dimensional graph where the vertical (*x*) axis represents the passage of time (labelled from 0 to 24 hours) and the horizontal (*y*) axis represents distance around your normal place of residence. To do this, you should first try to remember the activities (or projects) you conducted on any arbitrary day (e.g. yesterday). Write these down in the order that they occurred, noting the space and time that each occupied, then transfer them to your graph. In some cases – if you stayed in bed all day, for example – this will not be particularly difficult (or interesting), consisting of a single vertical line indicating the absence of movement. In most cases though, you will have moved between a number of stations and will be able to indicate this through a combination of vertical lines (representing periods spent in one station) and diagonal lines (indicating movement between them). Obviously, your 'scale' here will depend on the distances between the stations where these activities occurred, and it may be that it is more appropriate to consider using metres/feet rather than kilometres/miles if your activity space was restricted to a very localized set of stations. Once you have completed one day's path, you could try adding different coloured lines to indicate other days. As you do this, regularities may emerge. Considering these, try thinking about how your daily path and life-path are limited by your commitments to particular individuals and organizations (family, children, friends, lovers, employers) as well as your access to particular modes of transport. How could your mobility be improved?

One obvious point that is made through these examples is that people's life-paths are constrained on a variety of spatial scales (i.e. they are restricted in where they can and cannot go). The physical limits which mark the boundaries of the spaces that a person can access in the time available to them can be represented diagrammatically through **time–space prisms.** In the above example, the potential volume of space and time that is within reach of the car-user is much larger than that of a person who is reliant on walking, meaning that they have a larger space–time prism.

Overall, it is recognized that these spaces of possible activity are constrained by three major factors (Golledge and Stimson, 1997). The first of these are **capability constraints** – those factors which serve to limit the distances which a person can move in a given time-span. Some of these are the product of individual biology – the fact that some people are less able to move because of disability, age or illness (see Matthews and Vujakovic, 1995) – while others relate to the fact that some people have limited access to forms of private or public transport. Moreover, it is apparent that the need to allocate time for basic physiological necessities such as eating and sleeping also limits the time available for movement. In contrast, **coupling constraints** concern the fact that some projects involve the coming together of different individuals. Obviously some tasks (or projects) require a number of people to meet in the same place at the same time, involving the convergence of a number of individual paths on the same station. For example, this means that most people cannot simply choose when to go to work, but must be in the workplace at a time when other people who are involved in the production process are also there. Another example might be that a student is expected to turn up at university when a lecturer is giving a class rather than expecting that they can come in when they feel like it (or when they wake up). Finally, it is also important to consider **authority constraints**, which refers to the type of legal and quasi-legal controls which make certain spaces inaccessible to certain individuals. Obvious examples here relate to the way that children are not allowed in pubs (and are supposed to be in school for set times each day), the ways non-employees are excluded from certain workplaces and the way many shops, malls and public buildings are locked outside 'opening hours' (see also Chapter 8).

An example of how these different constraints combine to influence people's mobility is provided by Wilton (1996). His study of individuals living with HIV/AIDS suggests that this disease generally causes people to have a reduced daily path. On one level, this is because of the debilitating influences of the condition, whereby people's physical deterioration results in movement being impeded. Here, he cites the way that anti-viral medication may cause inflamation of joints and muscles, together with the onset of specific health problems, as a prime factor limiting mobility. Moreover, many of the individuals he spoke to acknowledged the psychological problems of living with HIV/AIDS, in that they spoke of the lack of confidence that they had in using public transport or going to places where they could be exposed to illnesses that might have serious consequences for them. Post-diagnosis, many referred to the difficulty of remaining positive and determining to live life as they had before, with the inevitable result that they were not as 'outgoing' as before. Yet these capability constraints on mobility were also accompanied by coupling constraints as these individuals found they needed to be in specific places at certain times to take medication, use health facilities or participate in support group meetings. Finally, authority constraints of various kinds were also limiting factors, with particular places recognized as off-limits because of the prejudices of particular communities and the occasionally negative reactions of employers, colleagues and even family members to HIV/AIDS. Collectively then, Wilton refers to the 'diminished world' of those living with HIV/AIDS, noting, however, that mobility following diagnosis proceeds through different phases of physical deterioration, social ostracization and emotional trauma (see Figure 2.5).

Undoubtedly, the time–space diagrams introduced by Hägerstrand and others have done much to alert geographers to the microgeographies of everyday life. Alan Pred (1985), for example, has contended that such diagrams are a mode of representation (see Chapter 7) that has the ability to make the geographies of everyday life visible while alerting us to the small details often hidden in the bigger (global) picture. However, others suggest that these skeletal diagrams fail to convey the way these geographies of everyday life are constructed by *humans*:

Figure 2.5 The diminished worlds of individuals living with HIV

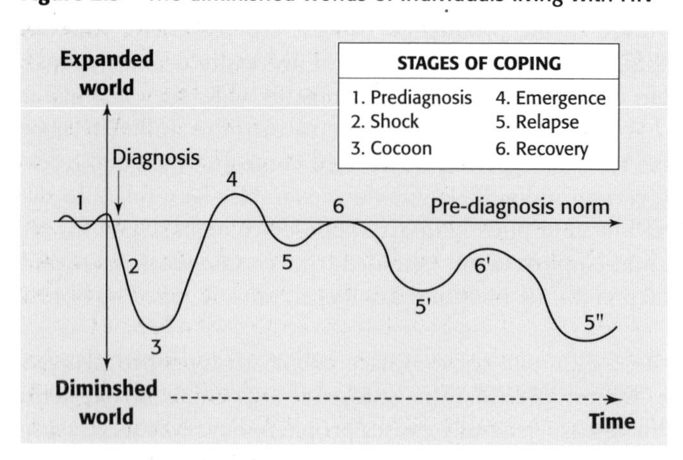

Source: After Wilton (1996)

> The fact that a human path in the time-geographic notation seems to represent nothing more than a point on the move should not lead us to forget that at its tip – as it were – in the persistent present, stands a living body-subject, endowed with memories, feelings, knowledge, information and goals – in other words capabilities too rich for any conceivable kind of symbolic representation but decisive for the direction of paths.
>
> (Giddens, 1984, 102)

While there are many concepts here that demand more careful consideration (What constitutes the body? What do we mean by feelings? What is knowledge?, etc.), a more basic criticism of time-geography emerges here, namely that it follows in the tradition of spatial science by dehumanizing what Alan Pred (1977) once called the 'choreography of existence'. So while the Lund approach began to alert geographers to the **spacing** and **timing** of human behaviour, it is important to remember that time–space prisms are not simply containers for human behaviours and activities; rather, they are activity spaces which are made meaningful and 'real' by people (e.g. through language, action and practice).

We might also start to think about the fact that 'stations' are not just points in space, but are **places** which are occupied, experienced and changed by human beings who are concerned with making a living, seeking enjoyment, making friends, etc. In short, these stations are places whose identities are negotiated in the midst of the complex personal and social interactions that occur there. An awareness of this is displayed in Pred's own work (as highlighted by Gregory, 1993). Concerned with documenting the geographies of nineteenth-century Stockholm, Pred (1990) uses a time–space diagram to reconstruct the daily path of Sörmlands-Nisse, an imaginary docker whose movements were taken to indicate the rhythmic motions which characterized this newly industrialized city (Figure 2.6). Yet alongside this, Pred emphasizes the importance of using other means of representing the docker's experiences in the city, trying to grasp the multiple ways that the docker understood his world and made it meaningful to himself. As such, Pred sought to use the bare bones of the line diagram as a basis for hanging other images and descriptions of Stockholm at the end of the nineteenth century – of the stench-drowned visit to the outhouse, the smell of stale beer and stale cigar smoke in the open-air cafés, people dozing on sacks of grain on the dockside, ships' flags whipping in the breeze and so on (Gregory, 1993).

In spite of these limitations, 'chronogeographical' analyses of the type pioneered by the Lund geographers are of interest as they focus on the individual, the short-term and the small-scale, pointing out the routine (and cyclical) nature of everyday life. By this stage, you may have noted that we have used the term 'everyday' on numerous occasions without really defining what is meant by it. To some extent, this is deliberate, as the everyday is something that is close and familiar to all of us, something that is invisible but ever-present (Roberts, 1999). In this sense, it is almost undefinable, being that realm of the routine and humdrum which we take for granted. It is, perhaps, understandable that the concept of the everyday has rarely

Figure 2.6 *Danse Macabre?* The daily path of Sörmlands-Nisse

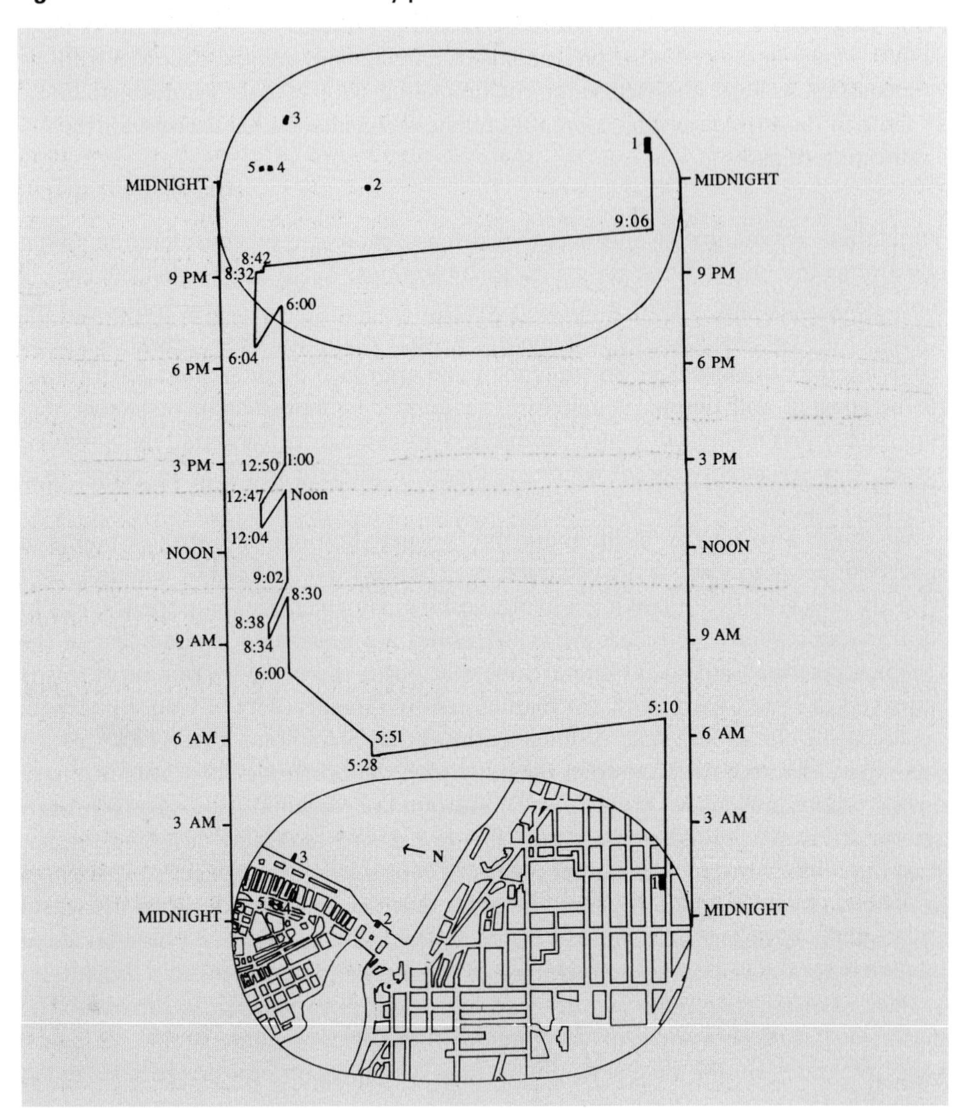

Notes: 1 = residence at *Bondegatan* 17; 2 = café at *Fiskarehamnen*; 3 = ship docked at *Skeppsbron*; 4 = café on *Österlånggatan*; 5 = public bar on *Österlånggatan*.

Source: Pred (1990)

been discussed or conceptualized in an explicit manner by academic theorists. A notable exception is in the work of Henri Lefebvre, a French philosopher whose work transcends history, geography and sociology. Lefebvre's interest in everyday life and 'lived space' occupied him throughout his life, including a three-volume 'critique' of everyday life which was published in 1947, 1962 and 1981 (with only the first volume having been translated into English – see Lefebvre, 1991). Disturbed that Marxist theorists seemingly downplayed the importance of lived experience in favour of discussions of high-level political ideas about capitalism, he sought to argue that it was in the realm of the everyday that the pervasive (alienating) effects of capitalism could be observed. As he put it in his own unique fashion, everyday life can be compared to fertile soil, and 'although a landscape without flowers or magnificent woods may be depressing for the passer-by, flowers should not make us forget the earth beneath, which has a secret life and richness all its own' (Lefebvre, 1991, 87). Here, he berated those theorists who focused on the spectacular 'flowers' of philosophical, artistic and political life as ignorant of the 'dirty soil' of everyday life. Implicitly, Lefebvre sought to attack those academics and theorists who sought to raise themselves above the 'mass' of ordinary people – those who sought to ascend the creative mountain tops by ignoring the plains and marshes of everyday life. Here then, we might see a connection between Lefebvre's concerns and those of the geographers who have argued for a human-centred geography (see Chapter 1).

Crucially, however, Lefebvre's concern for the everyday was based on the understanding that the everyday had been increasingly colonized by (global) capitalism, with the idiosyncratic lived spaces of (working-class) people being papered over by the **abstract space** of capitalism. While this argument is somewhat complex (and shot through with romantic nostalgia for the past), in re-reading his work it is possible to discern a genuine concern for everyday lived spaces as being where 'real' human life can be observed. This is expressed in his description of two contrasting (but typical) settings:

> Compare an 'average' house in one of our towns, not with an ostentatious and absurd palace, nor with some characteristically grotesque dwelling of the haute bourgeoisie, but rather with a 'modern' industrial installation – a power station for example. Here we find hyper-precise technology, light and a dazzling cleanliness; power condensed in strictly contoured appliances. . . . On the other hand, in the house where decent, 'average' people live out their daily lives, all is petty, disorganised, dusty nooks and crannies, mean, pretentious furniture, knick-knacks . . . .
>
> (Lefebvre, 1991, 231)

Elaborating this, he contends that it is here in the trivialities of the everyday that the essence of human existence can be discerned, and, accordingly, that there is nothing trivial about it at all. Or, as he put it elsewhere, 'Why wouldn't the concept of everydayness reveal the extraordinary in the ordinary?' (Lefebvre, 1991, 7).

For Lefebvre, then, it was essential to consider how individual people interact with other people in the realm of the everyday. Yet even here, the definition of the everyday remains slippery. Accordingly, Lefebvre sought to define everyday life principally in terms of what was 'left over' after all 'superior' political and artistic activities had been singled out, citing leisure, work and 'private life' as the essential components of the everyday. For him it was important not to separate these, as 'every week, Saturdays and Sundays are given over to leisure as regularly as day-to-day work' (Lefebvre, 1991, 30). Here he began to stress that what we are as human beings is not just determined by our role as workers, but also in our routinized 'private', family and social lives. In most of the urban West, therefore, we could suggest that the everyday refers to the customary and routine behaviours that occur in the workplace, the home and garden, the streets, shops, parks, cinemas, places of worship, football stadia, community centres and so on. To a lesser or greater extent, these are places where people adopt everyday modes of behaviour and thought, conforming to taken-for-granted assumptions about the way that people should act towards members of their family and their neighbourhood, apparently without even questioning this (but see Chapter 9). Against this, we might suggest that these habitual routines and behaviours are disturbed by the occurrences that punctuate the everyday – events that are, consequently, seen as 'extraordinary' when contrasted with the 'ordinariness' of everyday life. Examples here might include the rituals that accompany the special (religious) festivals, birthdays and carnivals that offer a break from humdrum routines. More exceptionally, catastrophic events such as wars, natural disasters and political insurrection may also disturb the everyday, exposing it as just that (although events such as war may unfortunately become considered 'everyday' in some contexts).

Accordingly, a geography of the everyday might simply be defined as concerned with the places in which everyday activities occur. As we began to see in Chapter 1, since the 1960s geographers have begun to interrogate the type of behaviours and lifestyles that are routinely found in particular locales, often using human-centred methods of enquiry to explore the specificities of people–place relations (see Chapter 10). However, it is apparent that many human geographers have neglected the everyday in their enthusiasm to document the exceptional, the new and the exotic (see Duncan, 1993). This can be illustrated with reference to a number of topics of geographical enquiry. For instance, while there is an extensive geographical literature on shopping as an activity central to people's lives, much of this currently concentrates on a few mega-malls (such as the West Edmonton Mall in Canada), overlooking routine acts such as the weekly grocery trip to Wal-Mart or Safeway or popping out to a local corner shop for a pint of milk and a packet of cigarettes. Equally, much writing on the geography of leisure and recreation examines the types of activities occurring in spectacular settings and theme parks such as Disneyland (or in the midst of events like the Olympics) rather than the leisure time which people spend watching television, or pottering in their garden.

In a thought-provoking essay, John Eyles (1989) follows similar arguments in suggesting that it is desirable that geographers examine the everyday. Furthermore, he contends that such a sensitivity to the commonplace is necessary for the development of a human-centred geography. This is a provocative argument (to which we will return in later chapters), but few could dispute that geographical analysis of the everyday offers a way of thinking about the relationships between individuals and their surroundings in a period of rapid time–space compression (see also May, 1996;

Watts, 1991). Specifically, in seeing the individual in relation to everyday life, one of the central questions of human geography emerges: what is the nature of the relationship between people and place? Inevitably, this question poses a series of difficult questions about how we might define place, about how we define people, and about the types of relationships that exist between the two. It is these questions that we shall seek to answer in the rest of this book.

■ Reading

Parkes and Thrift (1980) remains the key text on space–time geographies, while Chapter 8 of Golledge and Stimson (1997) is recommended as a summary of this literature which includes a more general consideration of time–budget studies. John Eyles' (1989) essay on 'The geography of everyday life' is of value as one of the only explicit attempts by a geographer to define the everyday, and while Miller and McHoul (1998) discuss the meaning of the everyday in relation to cultural studies, their chapters on the rituals of eating, shopping and conversation offer interesting interpretations of day-to-day activities.

■ Summary

In this chapter we have begun to make a case for exploring the everyday relationships that occur between people and place. Although definitions of the 'everyday' are open to interpretation, here we have stressed that a geographical interpretation of the 'routine' lives of 'ordinary' people is of interest for three interrelated reasons:

- While it is recognized that the very tissue of social life is becoming stretched due to the expanding reach of global technologies, media and transport, most people continue to lead lives that are based around a limited and localized array of places that are important in their lives.
- Although our lives are punctuated by special days or events (e.g. birthday celebrations, holidays, festivals), such events are the exceptions that punctuate the normal humdrum routines of work, rest and leisure that shape our lives.
- Although much geographical enquiry currently focuses on the networks, institutions and processes implicated in the production of global ways of life, focusing on the everyday encourages geographers to address the importance of people as more or less autonomous actors who creatively engage with, and shape, their surroundings.

three Knowing place

This chapter covers:

3.1 Introduction

In Chapter 2 we began to look at how people's everyday movements in (and between) places differ according to their social and bodily characteristics. In particular, ideas of space–time routines being structured around a limited number of places were considered, suggesting that people's lifestyles, while affected by global flows, are fundamentally distinctive and unique because of the individual and particular relationships they have with specific places. Yet simply to proclaim that everybody is shaped by their surroundings in different ways because they 'use' a different range of places is widely regarded as insufficient by geographers who have (generally) sought to elucidate the complex relationships that exist between people and place. In this regard, a major insight of human geography has been to indicate that different people may experience the same place in very different ways according to their **knowledge** of that place. As such, the relationship between people and place varies according to people's understanding of what happens in the place, how it has been designed, what its boundaries are and so on.

In this chapter, therefore, we want to examine a series of perspectives on people–place interaction that begins to problematize the rather straightforward conception of place that we have discussed so far. Specifically, we are interested in considering the distinctive contribution of what are generally known as **behavioural** perspectives to the understanding of people's relationships with place. As will be explained, such perspectives begin to undermine and rework the idea that places exist solely as real (or objective) phenomena which are experienced and understood in a similar manner by all individuals. Instead, behavioural perspectives alert us to the fact that each individual potentially possesses a unique understanding of their surroundings, and that this understanding is shaped by mental processes of information-gathering and organization. In this chapter, therefore, we explore the body of work by geographers that has focused on the complex ways that we obtain sensory information from, make sense of, and remember, places. In so doing, we will often be engaging with concepts derived from **psychology** that help explain how we perceive and understand our surroundings, showing how such concepts have been used by geographers to understand everyday behaviours in (and movements between) places.

In broad terms, psychology is the 'science of the mind'. Many people's image of psychology is that of a laboratory-based discipline, of scientists in white coats monitoring the behaviour of rats in mazes; for others, it might be of the psychoanalyst asking a patient to lie back on the couch and to relate childhood experiences. Either way, it might seem a little surprising that some geographers have looked to psychology to provide them with clues as to how people relate to their surroundings. Yet in this chapter we will see how a particular group of geographers has tried to develop (and continue to practise) a human-centred human geography using psychological concepts. Attempting to identify when this interest in psychology became apparent is by no means straightforward (Goodey and Gold, 1985), although certain figures have been recognized as particularly influential in expanding the horizons of geography beyond the realms of simple spatial analysis by incorporating psychological ideas. Gilbert White, William Kirk, John Wright and David Lowenthal have all been credited with bringing such ideas into the geographical fold, although further archival analysis reveals less obvious lines of intellectual heritage from the work of Carl Sauer and the Berkeley school (a North American group of historical-cultural geographers interested in the relationships between humans and environments in specific regions) and even the 'psychogeographers' who sought to develop a critique of urban life by exploring the effects of urban environments on behaviour and emotion (see Debord, 1967). Although many of these individuals and groups were writing in the 1940s and 1950s, their influence was primarily felt in the 1960s as dissatisfaction with the mechanistic and deterministic nature of the models prominent in the discipline began to take hold (see Chapter 1).

Within an intellectual climate which encouraged cross-disciplinary research, geographers sometimes found themselves working alongside psychologists who were simultaneously becoming increasingly interested in the influence of the physical properties of the environment on people's behaviour. As one of the pioneers of 'environmental psychology' recounted:

> Surprisingly it did not occur to psychologists that perhaps part of the variation involving (psychological) phenomena could be attributed to the nature, meaning, design, organisation and use of physical space. Because physical space was never made part of the problem, it could never be made part of the solution.
>
> (Proshansky, 1976, 305)

Accordingly, Harold Proshansky and other psychologists began to recognize that there were many ideas that needed exploring in terms of the relationships between environment and behaviour, with psychologists increasingly 'borrowing' (and reworking) geographical concepts of place and space. This was particularly evident in the pioneering work of Barker and Lewin on so-called **ecological psychology**, which sought to understand how different settings encouraged distinct behaviours (see section 3.4), as well as research examining people's visual perception of the physical environment. Geographers soon began to form useful cross-disciplinary links with many of these psychologists, founding new cross-disciplinary journals such as *Environment and Behavior* (1969). By the 1970s, the idea that geographers could profitably integrate psychological ideas into their work was well established, and in the remainder of this chapter we begin to explore some of these ideas – ideas that have led many to explore 'geographies of the mind'.

■ Reading

John Gold (1992) offers an interesting, reflective account of the origins and diffusion of behavioural approaches into 'mainstream' human geography. You might like to contrast his 'story' of behavioural geography with that offered in Cloke *et al.* (1991, 66–9) or Johnston (1991, 136–58). The differences evident here should emphasize that there are many different ways of writing geography, with attempts to 'pigeonhole' particular forms of knowledge into easy categories like 'behavioural geography' frequently glossing over the 'messiness' of geographical scholarship.

■ 3.2 Geographies of the mind, geographies of the senses

The need to know about the surrounding world is obviously one of the most fundamental human needs – being key to our survival on a day-to-day basis – but one of the most important ideas outlined by psychologists is that there is too much information in the world to take in at any given time. To comprehend this, it is important to realize that our interaction with (and hence understanding of) the world is constructed through our **senses** – principally sight, hearing, smell, taste and touch. Each of these senses is connected with specific 'perceptual receptors' which work to gather information from our surroundings and communicate it to our brains. For example, the photo-receptors on the retina at the back of the eye are responsible for conveying visual information to the brain, while mechano-receptors in the ear are responsible for communicating sound. For most people, these senses generally work together to provide us with clues about the places through which our body is passing at any one time. Each sense appears to offer characteristic ways of capturing environmental information, working on different scales with different efficiency. Taste, touch and smell are characteristically described as intimate senses, associated with immediate body space, while hearing and sight are represented as distant senses, able to capture information well beyond the body's immediate reach (Figure 3.1).

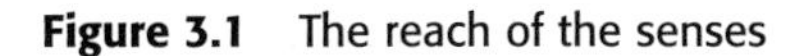

Figure 3.1 The reach of the senses

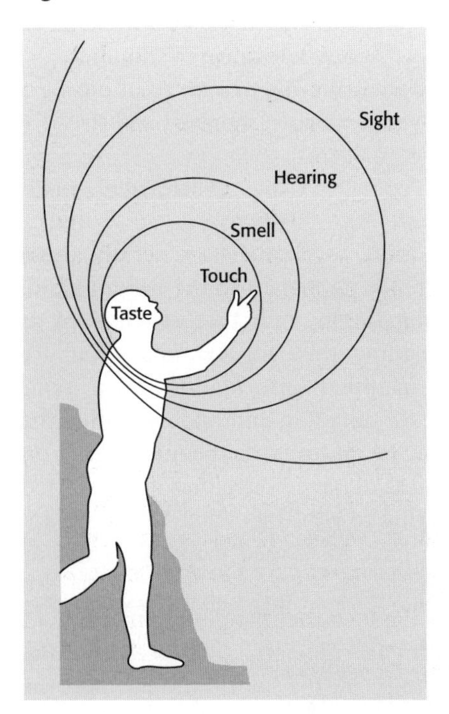

Source: Skurnik and George (1967, 14)

In fact, closer examination suggests that each sense works over both immediate and distanced ranges; for instance, it is possible for our sense of 'touch' to detect vibrations caused by distant objects, and our sense of smell may also be able to detect the source of particular odours over long ranges (Rodaway, 1994).

Crucially, though, the amount of information which a person can cope with at any one time is limited by the brain's ability to process the information that is being acquired. This is not necessarily an easy concept to grasp, but what psychologists generally contend is happening is that the brain often becomes 'overloaded' by the amount of information with which it has to deal. Even in seemingly 'simple' environments we are receiving a constant stream of environmental information via our senses (i.e. sights, sounds, smells, tastes and textures), and in busy, complex places (such as the centre of a major city) we are literally bombarded with a range of information, some of which will be unfamiliar to us, some very familiar. For example, city centres may be relatively quiet in the early morning, but:

> By eight or nine in the morning, however, cities take on a rather different feel and different set of associations. Among the more obvious is that the noise levels rise, as different kinds of sounds, many of them generated by the increased flow of traffic and people, distort and blend into one another. The result in many cases,

> is an almost undistinguishable cacophony of sounds . . . alongside the rise and fall in intensity of noise, however, is the wearily predictable smell of exhaust fumes, much of it diesel vapour, that assaults the nose . . . . A whole range of stimuli is there to confront and envelop pedestrians, from the scurrying to and fro of office workers . . . to the bustle of the streets themselves as tourists, shoppers and street vendors jostle for position.
>
> (Allen, 1999, 58–9)

To deal with this excess or overload of information, psychologists generally argue that we cope by 'filtering out' the information that is important to the task in hand, ignoring extraneous sources of information and gathering only that which helps us make pressing decisions. Even as you are reading this, for example, you are probably aware that your ability to take in other information from your environment has diminished as you concentrate on taking in and understanding the meaning of what is written here – so, you might be ignoring the people talking on the other side of the library, or you may have failed to have notice that it has started to rain outside. On the other hand, if you're trying to read this while watching TV or talking to a friend, you might find that you have to re-read many sentences as you haven't taken in what you've been reading!

Fundamental to a behavioural perspective is the idea that people's knowledge of their surroundings is perceived through the senses and mediated by processes of the human mind. The idea that people's behaviour in the world might best be understood by focusing on their **perception** of the world is often claimed to have been introduced to geography by William Kirk (1963, 365) who sought to make a distinction between the *objective* (or real) and *behavioural* environments. In his view, while the former consisted of the physical world around us, the latter consisted of the 'psycho-physical field in which phenomenal facts are arranged into patterns or structures that acquire values in cultural contexts'. Kirk thus believed it was the behavioural, not the objective, environment that provided the basis for human behaviour and decision-making. In effect, this idea challenged the assertion that human responses to environmental stimuli are based on the environment as it really is, and instead proposed that these responses are based on the environment as it is perceived to be (see Figure 3.2). The implication here was that human beings do not make decisions based on full, accurate and objective information about what exists in the world, but on what our senses tell us exists and what our brain is capable of dealing with. According to Kirk, our daily interactions with our surroundings can only be understood in relation to the partial, distorted and simplified understanding that we have of our surroundings.

It is, therefore, important to realize that behavioural perspectives are not necessarily concerned with behaviour itself, or its geographical distribution, but instead argue that patterns of human activity are best understood by examining the perceptions which influence these activities (Gold, 1980). Moreover, by emphasizing that knowledge is constructed by individual human beings according to their own

Figure 3.2 The behavioural environment

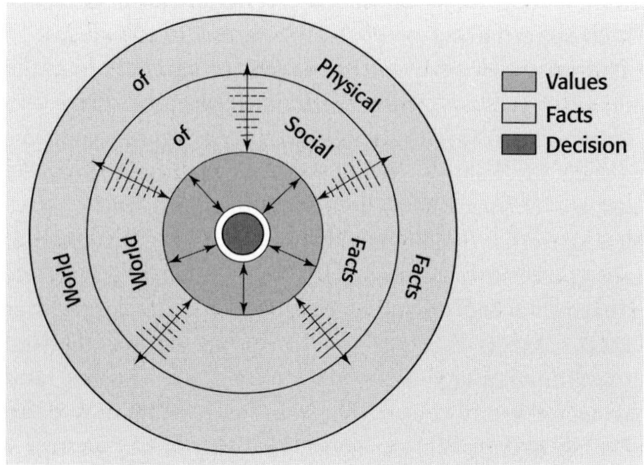

Source: After Kirk (1963)

subjective understanding of their surroundings, models such as Kirk's question the models of spatial science abundant in the 1950s/1960s, many of which suggested that people behave in a rational, objective fashion (see Chapter 1). At the same time, behavioural ideas seemed to explain why some models based on rational economic behaviour did not hold in the real world. For example, the idea that people did not have perfect information when deciding where to live, how to commute to work, or where to buy a pint of milk appeared to explain why some people's actual behaviour did not coincide with that predicted by geographical theories and models. This ushered in an understanding that human behaviour and decision-making is rarely rational, and is more usually **boundedly rational**. This term suggests that we seek to be rational within the limits of the knowledge that we have acquired about our surroundings, and that various options or possible choices are not known to us.

■ Exercise

One way of demonstrating the differences between the objective environment that 'really' exists and the behavioural environment which we perceive to exist is to think about notions of subjective and objective distance. For example, focusing on one routine activity, such as shopping, attempt to estimate the straight-line distance from where you are sitting as you read this to the nearest city centre, nearest supermarket, nearest convenience store and nearest post office. Then, use a map of your local area to measure the actual distances involved. What differences do you note? You could also try exploring other people's perceptions of distance. Are some people better at estimating distances than others? Why?

On examining your estimates of distance, you will probably find that you are some way out in most cases, and a long way out on some others. You might even find that your ideas about which supermarket or post office are actually closest to you were completely wrong. Thompson (1963), one of the first researchers to make a thorough investigation of subjective versus objective distance, noted that there was a widespread tendency for shoppers in San Francisco to overestimate distances between their home bases and their local shops. While you may well have done the same, other studies (some using travel time rather than straight-line distance) have revealed slightly more complex patterns. Specifically, Walmsley (1988) notes that subjective distances are usually greatly overestimated for short journeys, with the degree of overestimation decreasing as the objective distance becomes larger. But these findings are not universal, and it is often the case that people tend to overestimate distances for journeys away from town or city centres and underestimate the distance towards the city centre. The reasons for this may be to do with perceptions of travel times and possible congestion towards city centres – though they may also be connected to perceptions of the attractiveness of the shopping venue. In this case, we can perhaps see that our behaviour and decision-making are not informed by perfect knowledge, but a perception of the facts which results from our personal interactions with our surroundings as well as the information which we derive from **representations** of those surroundings. Of course, we may regard certain sources of information as more reliable and credible than others, inevitably treating a map showing a store's location as more 'accurate' than a description of how to get there given by a friend or neighbour (the way that we prioritize certain representations over others is further explored in Chapter 7).

These findings are of interest in so much as they begin to provide insight into why individuals go shopping in certain places (and not others); why they take certain routes (and not others) and whether they can even be bothered to go out shopping in the first place. As such, behavioural approaches acknowledge, explicitly or otherwise, that human action is mediated through partial and selective knowledge (i.e. based on particular perceptions of given situations or places). Strictly though, the use of the term 'perception' to describe the process whereby we understand our world is not correct, and models such as Kirk's were later amended by geographers concerned to distinguish between perception and **cognition**. In Kirk's model, the human being is simply imagined (and represented) as a passive decision-maker who acts on the basis of a world that they perceive through the senses and interpret through the mind, but the mind itself was left unexplored, a mysterious 'black box' which responded to certain types of information in particular ways. As a result, criticisms were widespread that simple behavioural concepts of a 'mismatch' between the objective and subjective world did not elucidate the role of human beings as active interpreters of their surroundings. As such, **cognitive models** were thus developed that began to focus more on the interpretation, synthesis and analysis of perceived information, recognizing that environmental stimuli merely represent inputs which need to be processed by the brain, with each individual dealing with this information in a different way.

One influential representation of this process was offered by Downs (1970) in his model of environmental cognition (see Figure 3.3). In this diagram, boxes represent concepts, and the arrows represent imagined links between them. While oversimplified, this diagram attempts to show that the real world is the source of information which is then 'filtered' through the five main perceptual receptors. In turn, this information is made sense of according to an individual's value system. Here, this is taken to refer to the attitudes and beliefs held by an individual (which are related to their age, gender, income, education and so on). This processing of information ultimately leads us to develop an image which, while partial, distorted and simplified, is useful in our day-to-day behaviour and acts as a basis for decisions and behaviours in place. Important here was the idea that not only do we all perceive different things as we interact with our surroundings, but that we creatively build this information up into images according to our individual psychological make-up *and* our role as members of different social groups.

Before we turn to explore the nature of these images, however, it is important to return briefly to questions of perception, as it is evident that many behavioural geographers have said little about the way in which we acquire information from the environment via our senses. As Gillian Rose (1993) contends, geography has always been a very visual discipline, prioritizing the gaze as a source of information about the environment and neglecting the non-visual aspects of human experience. Many of the 'fundamentals' of the discipline (like fieldwork sketching and mapping) are based on the idea that observation can capture most of what is important about a place, and that visual representation can be used to communicate this to others (see Chapter 7). But our images of place are not just built up from visual cues, and our knowledge of the environment also includes crucial information derived from taste, touch, smell and hearing. As such, much of the information we have about the environment is visual (i.e. the colour, texture, shape and arrangement of objects around us) but other senses contribute to our understanding of what surrounds us. Furthermore, the importance of sight relative to touch, taste, hearing and smell

Figure 3.3 A model of environment and behaviour

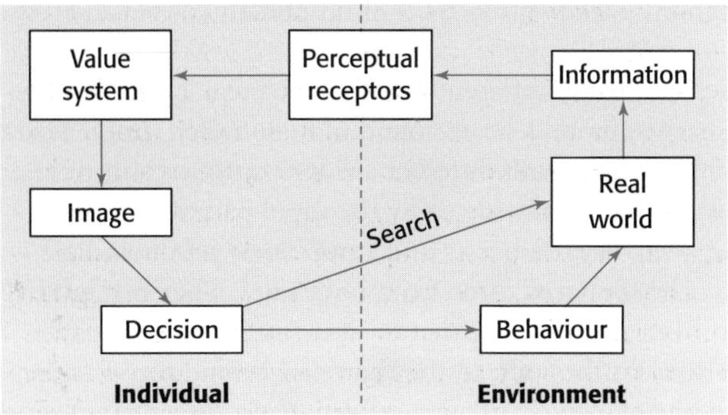

Source: After Downs (1970)

varies according to our physical condition, with the performance of our visual sensory apparatus influenced by factors of age, health and (dis)ability.

This point is, perhaps, reinforced by considering the way that those with visual impairment perceive and understand the world. Here, we need to be careful about generalizing across different severities of impairment, as visual disabilities vary from short-sightedness or a squint to total blindness. For the totally blind, however, many of the adjectives or adverbs we use to describe places (i.e. 'the red building', 'the first house to the left of the park', 'the shop with the big neon sign', etc.) may have no meaning. Equally, the comprehension of how places are spatially related to one another may be very difficult for blind people, with only patterns that are auditorily or tactically discernible being readily understood. This leads to what has been termed **perceptual inefficiency**, whereby visually impaired people are able to acquire information and make sense of their place in the world in a less 'efficient' manner than is possible for a sighted person. For instance, a blind person may be able to know and learn a route between two places, but may not be aware of a shorter route between them (though this may of course be a problem for sighted people too). Golledge (1993) thus contends that vision is of major importance when acquiring information about the environment, and that those who have visual impairment face major handicaps in their day-to-day activity.

For blind people, the environment is perceived through senses other than vision, particularly touch and hearing, potentially making these senses more acute. While sound and touch can never entirely substitute for loss of vision, these are used by visually impaired people to build up information about their surroundings, helping them navigate and negotiate their surroundings. Interestingly, when interviewing visually impaired people in Leeds and Reading (UK), Butler and Bowlby (1997) found that many blind people stated that they felt sighted people undervalued the potential of their full range of senses. Compared with sighted people, for instance, the blind may be more acutely aware of the 'feel' of different surfaces (such as tarmac, concrete, grass) under their feet, perhaps using their cane to provide further tactile evidence. Likewise, differences in volume, pitch and intensity of sound are often picked up on by blind people as a clue as to what surrounds them, as this description of life in a Birmingham park written by a blind person conveys:

> I heard the footsteps of passers-by, many different kinds of footstep. There was the flip-flop of sandals and the sharper, more delicate sound of high-heeled shoes. There were groups of people walking together with different strides creating a sort of pattern, being overtaken now by one firm stride or by the rapid pad of a jogger. There were children, running away in little bursts and stopping to get on and off squeaky tricycles or scooters. The footsteps came from both sides. They met, mingled, separated again. From the next bench, there was a rustle of a newspaper and the murmur of conversation. Further out, to the right and behind me, was the car park. Cars were stopping and starting, arriving and departing, doors were being slammed . . . . In front of me was the lake. It was full of wild fowl. The ducks were

quacking, the geese were honking and other birds which I could not identify were calling and cranking.

(Hull, 1990, 102)

Blind people, therefore, often develop the ability to distinguish between a myriad of sounds which most of us routinely encounter and 'filter out', reading these sounds to gain an idea of the environment where that sound was encountered and using this to orient themselves in place. Of course, we should not assume that the perception of sound gives the blind a continuous or reliable knowledge of the places they are in, as some places are poor acoustic environments where different sounds are muddied together in a way that makes it very difficult to discern the pitch and timbre of particular noises or voices. Even interventions like the noise of the wind can change or mask sounds in some places, making the familiar potentially unfamiliar.

■ Exercise

Go to a public place you know well – a park, a library, a shopping centre – and select a vantage point which would allow you to get a good view of who was around and what they were doing. Then close your eyes and try to 'listen' what is going on. The temptation may be to open your eyes to confirm what is making a particular noise, but try to keep your eyes shut and listen for a while. What do you hear? The longer you listen, does it become easier to discern individual sounds? Or detect where they are coming from? Later, you might try to write down some of the sounds you experienced, although trying to represent sounds using words can be problematic (emphasizing that language too prioritizes vision over other senses).

If you are sighted, this exercise may have begun to give you an idea of how visually impaired people perceive the environment – after all, blind people do not develop 'better' hearing, but are simply forced to use this sense more. But it is important to remember that if you are a sighted person, you can always simply open your eyes again. Related to this, while you might have got an idea about how blind people can judge distance and movement through hearing, you probably got little sense of what it means to be a blind person in the world. As Butler and Bowlby (1997) found, blind people often feel self-consciously different – like outsiders in a sighted world – and often feel dependent, vulnerable and reliant on others. Thinking about the capability constraints we discussed in Chapter 2, the ability of blind people to move around is often curtailed by obstacles like curbs, road signs, uneven paving, stairs or protruding objects, and given their limited ability to hear, feel or smell such inanimate objects, it means they may rely on the guidance of others if they want to move around without pain, embarrassment or hindrance. Therefore, we have to remember that sensorily-impaired people's experience (as well as their perception) of space is often highly constrained, resulting in a dependence on others

‥ipate in many of the activities which sighted people may take for granted, ‥o amount of 'role-play' can convey this.

Reading

These ideas about perception, cognition and environment are extensively explored in Walmsley and Lewis (1993), while Rodaway (1994) offers a fascinating geography of the senses that places much emphasis on the active role of the senses in shaping our knowledge of the world. Conflicting perspectives on the geographies of disability may be found in Imrie (1996) and Golledge (1993).

3.3 Place images and mental maps

As we have begun to describe, environmental knowledge is acquired through our interactions with, and movements between, different places. These interactions may be first-hand, as when we acquire information from a place that physically surrounds us, or it may be indirect, as when we experience a place vicariously, through media representations, maps, atlases, videos, paintings and so on. What is important is that this information is processed, via mental processes of cognition, to form stable and learnt images of place, which are the basis for our everyday interactions with the environment. Earlier, we cited the example of a bustling city centre to make the point that our surroundings are often more complex than the sense we can make of them. Work in behavioural geography has therefore begun to suggest that to cope with this complexity we carry mental representations or place images around 'in our heads' so that we navigate through this complexity with ease. One way of thinking about these images has been to suggest that they are constructed and built up from our perception and cognition of the environment (meaning that they are sometimes also referred to as **cognitive maps** or **mental maps**). These mental maps summarize each individual's knowledge of their surroundings in a way that is useful to them and the type of relationship they have with their environment. As such, these maps will be partial (covering some areas, not others), simplified (including some environmental information, but not all) and distorted (based on the individual's subjective environment rather than the objective environment). What is also apparent is that people hold a number of mental maps of different places on different scales – we might begin to think of each individual having a 'mental atlas' where each page can be mentally referred to in a particular situation. Individuals therefore construct images of places at different scales (e.g. rooms, houses, neighbourhoods, cities, regions, nations, the entire globe) so that they can use these images to cope with ordinary and familiar situations. It is only when the environment is unfamiliar or extraordinary that we consciously update our mental atlas. Once sufficient information has been acquired and processed to enable a decision to be made, the search for information ceases (this coincides with the idea developed by cognitive psychologists that individuals are discriminatory and active selectors only of that information which is relevant to their present needs). Images are relatively stable, therefore, but change in relation to increasing experience and required behaviour. According to this **transactional** approach, change prompts an individual to search for and select new information while simultaneously using old information to predict how they ought to behave.

The idea that we constantly draw on (and modify) these environmental images to cope with the complexity of everyday environments was most famously developed by the planner Kevin Lynch, who was largely concerned to understand how people looked at and made sense of their surroundings to promote better urban design. His book, *The Image of the City* (1960), inspired a generation of geographers to explore the mental maps which people make of urban space. Lynch's initial work used interviews with 30 people about their image of the inner city of Boston, an exercise that he repeated in Los Angeles and Jersey City. These people were asked what came to their minds about these cities, to describe their distinctive elements, to recognize and locate various photographs and even to accompany Lynch around the city describing their feelings about specific locations. Yet the most innovative aspect of the work, and the one that inspired most imitators, was the use of what became known as the mental mapping technique. This deceptively simple technique required that individual respondents drew a sketch-map of the city. This sketch-map was subsequently taken by Lynch to reflect their overall image of the city, and although Lynch recognized that drawing was an unfamiliar act for some people (as opposed to talking) he believed that this was a better way of eliciting spatial ideas from respondents than simply using verbal methods.

Though intended as a small pilot study, Lynch developed important insights from his exploration of people's mental images. Most importantly, he concluded that some cities were more easily known and remembered than others because of their inherent **imageability**. Here, knowledge of a place was seen to relate to its imageability and the extent to which it made a strong mental impression on the individual. Imageability was subsequently described by him as the ease with which the parts of the city could be organized into a coherent whole: 'That quality in a physical object that gives it a high probability of evoking a strong image in a given observer. It is that shape, colour or arrangement that facilitates the making of vividly identified, powerfully structured, highly useful mental images of place . . . it may also be termed legibility' (Lynch, 1960, 9). Lynch asserted that the constituent objects which most frequently featured in people's mental maps could be described as **paths, edges, nodes, landmarks** and **districts**. The distinctions between these are not always clear and will vary from person to person, with some features potentially being an edge for one person but a path for others. Nonetheless, a memorable and imageable city would be one where the paths, edges, districts, nodes and landmarks are all clearly identifiable and positioned relative to one another. Here, paths refer to those channels of movement along which people move (i.e. streets, walkways, rail lines, canals); edges are linear breaks or boundaries (like walls, rivers, shorelines); districts are distinct places which have 'some recognizable character'; nodes are strategic meeting points (like junctions, squares or street corners) while landmarks are defined as physical objects or public symbols (see Chapter 4) which may be seen from various vantage points and distances (a tower, a flag or even a mountain). Following Lynch's example many geographers have been involved in exploring the constituent components of an individual's mental map (see Figure 3.4). In this figure the percentage of middle-class respondents who mentioned specific features in their maps are indicated, distinguishing between Paths, Edges, Nodes, Districts and Landmarks.

Figure 3.4 Components of mental maps

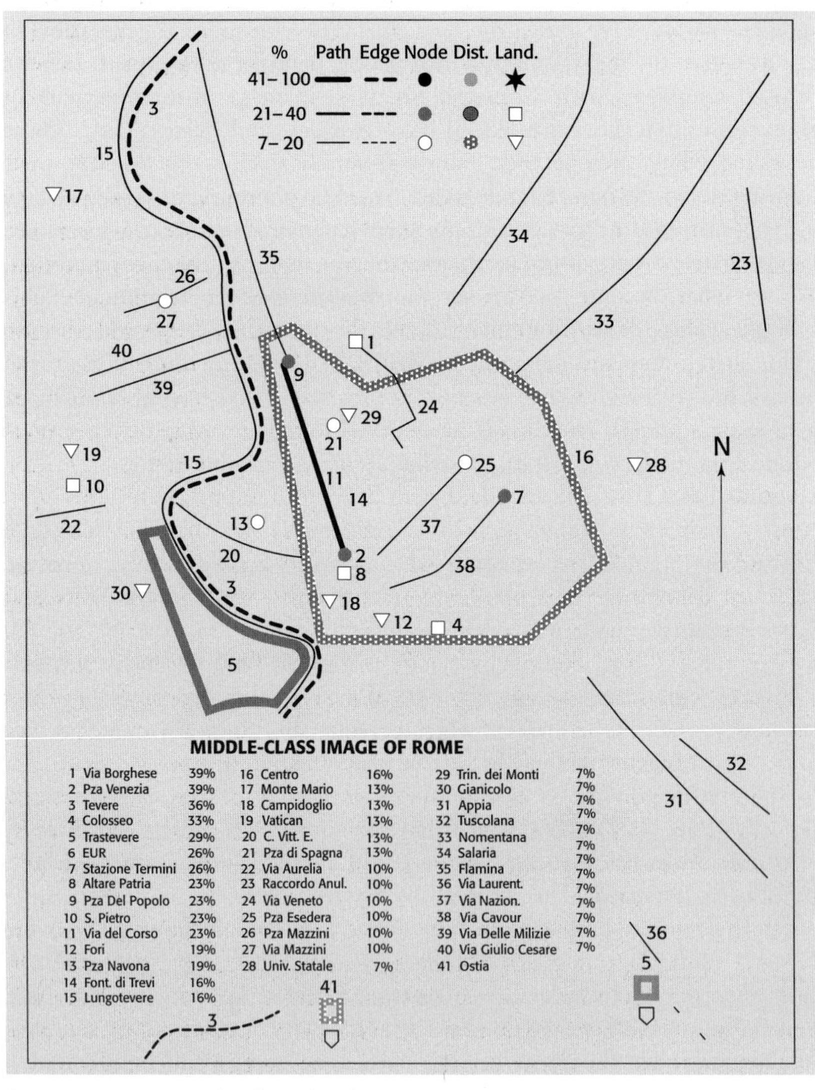

Source: Franscescato and Mebane (1973)

■ Exercise

Take a blank piece of paper and draw a sketch-map of a town or city you know well. Imagine you are drawing a sketch-map that would help someone who didn't know the town to find their way around – you shouldn't worry too much about orienting the map to north or following cartographic conventions. As you draw, you might like to think about how you know different areas, and how your knowledge of the town has built up (or perhaps diminished) over time. When you have finished, compare your map with a published map. What differences are evident? Are there districts which you have missed out? Distances which you have under-represented? Roads or paths whose route you have distorted? How might these simplifications or distortions shape your behaviours and movements in this place?

While pointing out the differences between mental maps and the 'real world' may be interesting, behavioural studies posit a mutually interacting relationship between the sketch-map which people draw and the type of mental images they carry in their heads. However, whether a flat, lifeless sketch-map can convey the nature of this relationship has been queried by a number of commentators. Golledge and Stimson (1997), for example, question whether a two-dimensional representation 'from above' captures how people make sense of the world, given that they navigate through space – generally on foot – experiencing it as a series of complex encounters which are viewed (and heard, smelt and touched) from the ground rather than from a bird's-eye perspective (this issue of how places are represented during mapping is one we return to in Chapter 7). One issue here is the question of whether a map can represent the smells, sounds and tastes which, as we discussed before, form an important part of our perception of place. Subsequently, both geographers and psychologists have drawn attention to the implicit limitations of mental maps for generating insights into how people make sense of the world. One major criticism centres on the idea that some people (like geographers!) are supposedly better at drawing maps than others because of their training and education (again, see Chapter 7).

As such, it is important to stress that the sketch-maps drawn by people are indicative of some aspects of their mental image and reveal some of the processes by which they make sense of their surroundings, but maps do not reveal how this information is actually carried around in the head. Indeed, the mental map is probably best considered as a metaphor or analogy, as psychologists are unsure whether the part of the brain where spatial information is processed (called the hippocampus) stores information in a spatially referenced form. Alongside sketching techniques, therefore, other methods have been used to try to ascertain information about the way in which people understand and organize information about their surroundings, including distance and direction estimation exercises, being asked to locate points on a base map, recognize an aerial photograph or verbally describe a route or area. On the basis of these exercises, some geographers have tried to develop specific understandings of how spatial knowledge is learned and recalled, elaborating on Lynch's claim that a person builds a comprehensive knowledge structure in which districts, nodes, edges, landmarks and routes are linked together to gain an understanding of the arrangement of things, people and events. Some of these ideas of spatial cognition are conceptually complex, and have yet to be integrated into an overall 'theory' of how we come to know place (though see Kitchin, 1994). Indeed, there are many debates as to whether we come to understand our surroundings through landmark-based learning (where we focus on the spatial arrangement of imageable landmarks) or procedural learning (where we focus on route-based knowledge). Nonetheless, most researchers agree that knowledge is hierarchically organized around major 'anchor points' (landmarks that are important in our day-to-day activities) upon which other information is 'hung', with paths often also acting as key reference points in our spatial knowledge.

Some clues to the way we come to know place can be acquired by studying children's development. Traditionally, children have been a neglected group in geographical enquiry, with few geographers interested in the relationships which children have with their surroundings (James, 1990; although see Ward, 1990a; 1990b). The fact that there has been much recent interest in children's geographies results, in part, from the idea that studying their spatial cognition is useful for appreciating the influence of both physiological (biological) and psychological development on the way information is perceived and processed. Studies in both laboratory and classroom settings have shown that children have a range of spatial cognition skills which, while different to those of adults, should not be underestimated. Mental mapping, drawing and role-playing exercises reveal that children not only represent their environment in a different manner than is the case with adults, but that this reflects different ways of understanding the world. Here insights from Piaget (a child psychologist), have been employed to highlight the fact that children do not simply acquire information about their surroundings in an incremental, linear fashion where bits of information are gradually pieced together over time. Rather, it appears that children develop through different phases of environmental understanding as their bodily and mental capacities change. According to Piaget, the child passes through a sequence of four discrete stages on the passage to maturity (see Table 3.1).

The crucial thing to note here is that each of Piaget's stages is associated with a particular way of understanding (and representing) the world. For instance, in the formal operational period, which normally lasts until about two years of age, we only have the ability to view and represent our surroundings from an **egocentric** (self-centred) perspective, lacking the ability to perceive it or represent it from another person's perspective. As we grow older, our ability to imagine the environment through more abstract frames of reference becomes obvious as we

Table 3.1 Children's understanding of their surroundings

	Environmental information (inputs)	**Cognitive processes**
Formal operational stage (0–2 years)	Exploration of own body. Object movement via motor actions (i.e. movements of body). Observation of object movements.	Development of body awareness. Relating object positions relative to self. Developing sense of magnitude/scale.
Preoperational stage (2–5 years)	All above plus images of place derived from stories, books, television. Extended sensory input resulting from increased personal mobility. Beginning of place exploration via play.	Beginning of ability to represent own place in the world. Distinction of general and particular object forms. Development of spatial referencing systems. Association of particular places with particular social activities and behaviours.
Concrete operational period (5–11 years)	Extended mobility beyond home into wider range of spaces. Involvement in forms of play and sociality in formal and informal settings. Varied experiences of travel. Increased exposure to images of place via formal education system.	Greater detail and precision of mental maps, capable of giving distance and direction estimates. Development of a diversified range of place images. Increased understanding of social rules of place.
Formal operational period (11 years onwards)	Individual mobility and travel generally free from constraint. Increased exposure to images of distant places via formal education and individual interests. Development of personal place preferences and tastes.	Generally precise mental maps covering extensive area, though subject to individual variation. Greater knowledge of distant places. Ability to apply previously learnt knowledge to new situations.

Source: Adapted from Walmsley (1988) and Spencer (1991)

begin to adopt a fully co-ordinated, spatial referencing system. There is some evidence that this is reflected in mapping ability as individuals are able firstly to draw egocentric pictorial 'maps' (centred on the self), then objective pictorial plans (indicating relations between objects) and finally abstract, spatially referenced plans (see Figure 3.5). Yet there is much variation observable in terms of what children know and understand at different ages according to their gender, class, income and education. For instance, Matthews (1992) found considerable differences in the detail and extent of spatial knowledge displayed by girls and

Figure 3.5 Children's mental maps

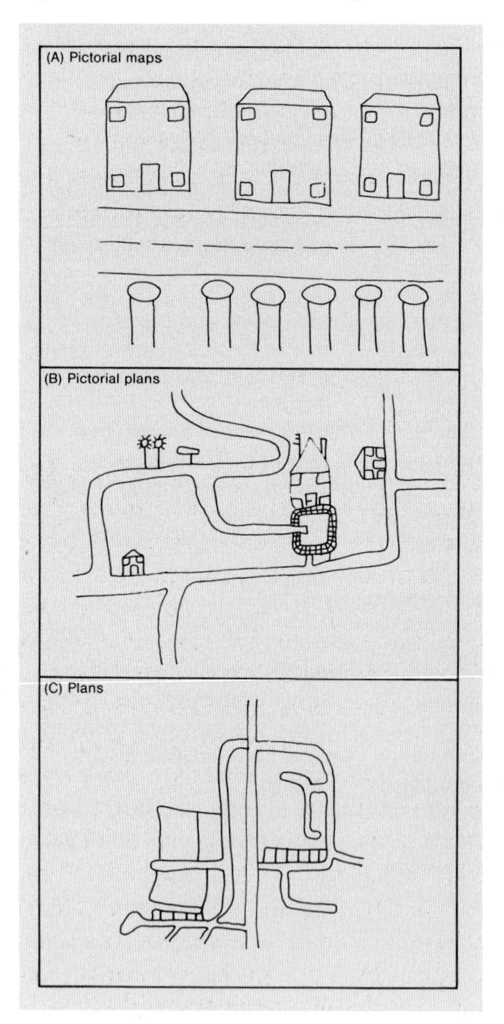

Source: Matthews (1992)

boys, with boys at one Coventry school able to draw much more extensive and precise maps than girls at the age of eleven. Here, parental influence may have been an important intermediate factor, as the girls were generally more constrained than the boys in terms of the type of places and images to which they were exposed. On a basic level, this observation reminds us that the acquisition and processing of environmental information varies between people according to their individual psychological make-up and their role as members of different social groups.

■ Reading

Golledge and Stimson (1997) provide a useful overview of theories and ideas developed in relation to spatial cognition and the development of spatial knowledge. The coverage of the book is extensive, so you might want to focus on Chapters 5 and 7 to gain further insights into how we come to know place. Matthews (1992), summarizing a number of his studies integrating geographical and psychological ideas, explores how children make sense of their surroundings, while Aitken (1992a and 1992b) provides a more extensive consideration of different theories of environmental cognition and learning.

■ 3.4 Behaviour in place

We have perhaps given the impression so far that behavioural approaches are solely concerned with the way in which people learn the arrangement and 'geometry' of places. Certainly, behavioural perspectives appear to offer major insights into the way people's decisions to move between particular places following particular routes are influenced by their own partial, simplified and distorted view of the world. However, ideas of perception and cognition have also been used in attempts to understand how people *behave* in particular settings. For instance, the idea that some areas are perceived and understood as places where certain behaviours are acceptable or normal has been seized upon by geographers exploring the relationships between crime and place, and survey and questionnaire work with convicted criminals has begun to indicate how their knowledge of space informs their decision to commit crimes in certain areas – whether these crimes are burglaries, thefts of vehicles or even rapes (Brantingham and Brantingham, 1992). Some recent studies have highlighted that environmental stimuli (or information) might be understood by potential criminals as indicating an area where a crime may be carried out with only a small risk of being caught. Litter, graffiti, vandalism and poor maintenance of open spaces (in particular) might all indicate that a place is not looked after, suggesting that it is more vulnerable to crime. While this is not to imply that these **environmental cues** *cause* crime and criminality (after all, criminals must be motivated to commit crime in the first place) it does suggest that people's perception of place is crucial in shaping the way they behave in that place (Herbert, 1993).

This link between place and behaviour has also provided a major research focus in environmental psychology. Specifically, many studies have explored the way individuals behave in clearly defined (and often enclosed) places referred to as **behaviour-settings**. According to one of the pioneers of such research, Roger Barker, a behaviour-setting is a bounded space that is constructed and defined through two sets of components, 'psychological' and 'non-psychological'. Here, the former consists of specific forms of behaviour and action while the latter consists of the physical objects with which that behaviour is carried out. The 'unity' of the behaviour-setting is seen to derive from the interdependence between these two components, with the events occurring within the boundaries of the behaviour-settings seen as having a greater effect on one another than events outside its boundary. To examine the significance of behaviour-settings in people's lives, Barker and his colleague Herbert Wright established the Midwest Psychological Field Station in the small town of Oskaloosa, Kansas, in 1947. Designed to operate as a centre from which people's behaviour in the town could be observed, the station allowed Barker and Wright to examine how specific 'real' environments were perceived and used by people in the context of their everyday lives (Bonnes and Secchiaroli, 1995). At the time, this attempt to take psychological research out of the laboratory was seen as revolutionary within a discipline underpinned by a belief in the power of experimental, scientific method, yet this approach was to pre-empt many of the concerns of later psychologists that their research should have practical significance and application.

Based on painstaking and meticulous observation of people's behaviour (at a time when video-recording techniques were not available), Barker was concerned with describing and explaining the occurrence of the particular behaviour episodes which occurred in specific behaviour-settings. For him, behaviour-settings were 'the basic facts of life for the people in the town' (Barker, 1968, 51), with different ritualized, taken-for-granted behaviours evident in the town's grocery store, its school, its beauty parlour, its Rotary Club, and so on. The example of the town's Presbyterian church was often cited by Barker as illustrating the importance of behaviour-settings in structuring social life; the layout of its benches, altar, pulpit and entrance were all seen as fundamental in shaping how the setting was used in religious festivals and celebrations. Moreover, Barker began to speculate that slight changes in this layout could disturb established patterns of behaviour. With its reliance on fastidious and close description of people's relationships and behaviours in specific settings, Barker's work had strong parallels with the in-depth studies of urban life pioneered by sociologists from the Chicago School. However, Barker's interest in behaviour settings was motivated by a strong interest in exploring how settings might be modified to ensure an effective match between the function and form of the behaviour-setting.

Following his work in Oskaloosa, for example, he began to focus on one of the behaviour-settings that he felt was most influential in society – that of the classroom. For him, the classroom was a clearly identifiable (i.e. discrete) unit observable within the 'ecological environment' characterized by a high level of internal interdependence:

> There is a carefully negotiated balance between the patterns of behaviour happening in the class and the pattern of its non-behaviour components, the behaviour objects. The seats face the teacher, for example, and the children face the teacher . . . . A pupil has two positions in class; first, he [*sic*] is a component of the supra-individual unit, and, second, he is an individual whose life-space is partly formed within the constraints imposed by the very entity of which he is a part.
>
> (Barker, 1968, 95)

From this perspective, Barker began to argue that the balance between the physical objects that constitute the classroom (including the pupils themselves) and the learning behaviour supposed to occur in that classroom can easily be thrown off balance. One way in which this might occur, he suggested, is by having the wrong number of pupils present in a specific classroom; assuming that each classroom is designed for an optimum number of pupils, Barker referred to too large classes as 'overmanned' [*sic*] and too small classes as 'undermanned'. Exploring this contention in schools around the USA, he used measures of educational attainment and satisfaction to demonstrate that when classrooms were occupied by more pupils than they had been designed for, participation and attainment levels dropped. When the number of pupils was fewer than the optimum, their participation and cohesion was seen to increase while the effectiveness of the teacher was often seen to decrease.

Although Barker's work on the importance of class size did not pay much attention to the type of social relations being played out in the classroom (see section 8.1), his work inspired much subsequent behavioural research, some of which has proved influential in educational debates (for a summary, see Weinstein, 1979). Alongside such research, however, has been another important body of research which has ignored the influence of class size *per se* to focus on the physical design of classroom environments. Such research, based on psychological experimentation and **post-occupancy evaluation** techniques (where pupils and teachers complete questionnaire surveys to indicate their satisfaction with a teaching environment), has suggested that there is a significant relationship between design characteristics and educational performance. For example, particular design features have been shown to increase interaction between lecturers and students, promote friendships between students as well as diffusing aggression. Indeed, one of the most widely cited examples of successful classroom modification was developed by a psychologist who conducted a series of collaborative experiments at the University of California involving university administrators, designers and psychologists. The impetus for this study came from Robert Sommer's (1969) own teaching experiences, which were characterized by low levels of student participation despite his best efforts to include and encourage discussion. Questioning his own ability as a teacher, Sommer observed other teaching sessions and found his own experiences replicated across his faculty; even in small seminars, he found student contributions occupied an average of just under 6 minutes per hour. Considering this lack of interaction, Sommer began to speculate that the traditional design of classroom, with desks arranged in straight rows facing ahead, discouraged student discussion and meant that interaction was limited to that between the teacher and the students. After many tests and years of experimentation, Sommer rejected this **sociofugal** design (one that discourages interaction between individuals) in favour of a **sociopetal** one (that encourages interaction) where seats face inwards in a relaxed, carpeted 'Soft Classroom' (Figures 3.6 and 3.7). The fact that the soft classroom design has been adopted in a variety of contexts across the world seemingly indicates that psychological theories can be harnessed to solve particular design problems.

The insights of environment-behaviour studies have subsequently also informed the design of office settings, sports stadia, restaurants and bars, where particular design features are seen to encourage more sociality and co-operation between people (Baldry, 1999). Of course, the type of design change recommended by the researcher depends on the type of behaviour which the setting is supposed to encourage. For example, the idea that we all desire an amount of **personal space** has been used to inform the layout of seats or reading booths in libraries, where it is often the case that social interaction (particularly talking) is discouraged. Here, the idea that people seek to avoid personal contact with others in an imaginary 'space bubble' that surrounds our body at a distance of about four feet (see Figure 3.8) has made designers sensitive to the way in which people use library reading rooms. As Eastman (1975) found, users of libraries generally seek to find a seat at an unoccupied table, avoiding seats within the personal space of other users. Within Western societies, it is argued that this is because people feel crowded and anxious if they have to sit or stand closer than this to any person with whom they do not have an intimate relationship. As a result, many libraries set aside areas where seating around tables has been abandoned in favour of privatized carrels or work-booths, which are seen to represent a much more efficient use of space.

In the supermarket, too, insights from environmental and consumer psychology have been harnessed to shape people's shopping behaviour. Many of the 'tricks' which supermarkets use to entice people to spend more money, and increase their 'dwell time' in the shop, are now well-known. Environmental modifications like the use of bright lighting to enhance the appearance of fresh fruit and vegetables, pumping the smell of freshly baked bread around the store or playing calming music are common ploys designed to make customers spend longer shopping, while diffusing the stress often associated with shopping. As competition between major superstores has become intense, this range of techniques has expanded in an attempt to influence customers' movement patterns around the store – for instance, interior

Figure 3.6 A 'hard' classroom

Photo: Authors

lighting may be stronger at the rear of the store to 'draw' people in; everyday essentials like milk and bread may be placed strategically to ensure customers have to walk past other tempting offers first. Consumer psychology has even suggested to retailers that it might be beneficial to position their own label brands to the left of

Figure 3.7 A 'soft' classroom

Photo: Authors

more recognized names if they want people to consider buying them – the reason being that most people 'read' the shelves from left to right (Figure 3.9). Through such strategies, retailers hope to promote particular goods over others, trying to persuade us to make impulsive and spontaneous purchases ('splurchases' in the

Figure 3.8 The notion of personal space

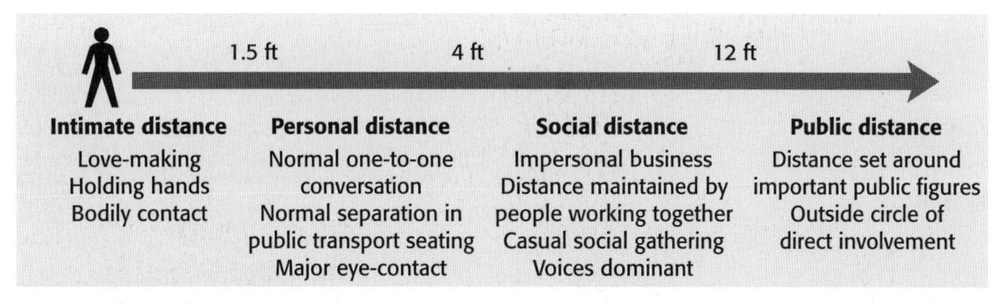

Source: Based on Sommer (1969)

trade jargon). Some retail analysts have even blamed the falling profits of some retail chains on the fact that they have not spent enough time and money thinking about store design and layout, with Woods (1999) claiming that established Marks & Spencer's clothing stores have not been able to keep up with newer competitors (such as The Gap) because they fail to provide the right ambience for clothes shopping.

■ Reading

There are many textbooks exploring the links between environment and behaviour. Many are written primarily for psychology students, but are widely accessible. Try dipping into Bechtel (1997) or Cassidy (1997) and you'll find some interesting case-studies of how designers and researchers have worked together to 'manipulate' behaviour in different places employing ideas from personal space and crowding theory. The journals *Environment and Behavior* and the *Journal of Environmental Psychology* are also widely available, containing case-studies of design intervention alongside more theoretical pieces on the psychology of place.

In all these examples, the idea that behaviour is shaped by the physical environment and design of place is one that comes across strongly (echoing the ideas of environmental determinism implicit in the regional geographies discussed in Chapter 1). However, later refinements of this idea by environmental psychologists have sought to undermine this rather straightforward reading of the relationship between people and place. For the psychologist David Canter, for instance, the idea that people arrive in a place and find themselves acting in a particular way according to their perception of its physical characteristics appeared ridiculously simplistic. Refuting this idea, Canter (1977; 1991) suggested that it was crucial to investigate **purposive behaviour**, arguing that the reason (or reasons) that someone is in a place acts as a major influence on how that place will shape the person's behaviour. By way of example, he argued that a restaurant isn't simply a place where the layout of

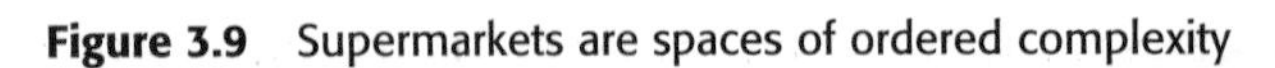

Figure 3.9 Supermarkets are spaces of ordered complexity

Photo: C. Rowley

tables, seats and counters cause eating to occur, but that it is a place where people go to eat and that this act is made more or less easy, pleasant or leisurely depending on its layout and design. In this sense, he argued that a restaurant would be experienced and evaluated very differently if it was being considered as a place to hold a lecture as opposed as a place designed for eating in. Canter therefore suggests that understanding the experiences that relate a person to a place and explain the person's reason for being there are the key to exploring behaviours in place. Of particular interest to Canter was the question of someone's **environmental role**, defined as that aspect of a person's social role which is related to their dealings with the immediate physical surroundings (Canter, 1991). In relation to the example above, it might be suggested that a waiter and a diner would experience the same restaurant very differently, and that some changes to its design would inevitably affect waiters more than customers.

Developing this point, Canter describes the act of eating out – whether in a *haute-cuisine* or fast-food restaurant – as being constrained by place rules, norms and codes. Here, he is thinking not just of the laws that dictate at what age you can consume alcohol on the premises, or whether you can smoke in a particular area (see Chapter 8), but more general perceptions of what behaviour is acceptable in the place. Again, environmental cues and design features may be crucial for conveying ideas that certain forms of behaviour are expected there; for instance, the perceived quality of decor, furniture and presentation of food may all lead to the understanding that particular forms of etiquette need to be abided by. Both those who eat and those who serve are subject to these influences, with the setting of the restaurant designed to create a suitable atmosphere for the rituals of eating out to be played out (see also Bell and Valentine, 1997). Part of the task facing restaurant designers is, then, to provide a design which effectively meets the needs and wants of specific user groups, creating the right sort of ambience for the type of clients that they are seeking to attract, encouraging more affluent groups to linger and indulge in a *haute-cuisine* restaurant while poorer groups might be discouraged from lingering in fast-food restaurants by a functional, sociofugal layout of seats and tables (Shelton, 1990). The fact some people may not understand the rituals and rules associated with particular places (e.g. expecting table service in a burger joint) suggests that behaviour settings are experienced differently by different individuals; while some may interpret the design as encouraging a particular form of eating, others may not. As such, while some may feel 'in place', others are 'out of place' (a point we return to in later chapters).

In alerting psychologists to the importance of people's social role and the importance of place rules, Canter and others began to demonstrate that many psychological explanations of people's behaviours in place were based on a limited understanding of people–place relationships. In particular, by highlighting the contingency and variation in individual responses to different settings, recent work in environmental psychology has sought to stress that it is important to take a **behavioural** perspective on people–place relations rather than one that is

behaviourist. The distinction between these two concepts is more than a matter of mere semantics with both representing fundamentally different schools of thought about the nature of the relations between place and behaviour. Although there is no one single behaviourist approach, Walmsley and Lewis (1993) argue that it encompasses a belief that only those behaviours and objects that can be physically **observed** and **measured** have any legitimacy in the study of the effects of place on behaviour. As such, behaviourist approaches tend to reduce complex interactions between people and places to simple stimulus-response models where the real environment is seen to shape an individual's behaviour in a more or less deterministic fashion. In clinical and experimental psychology, such models have often been developed on the basis of experiments using laboratory rats, leading critics of behaviourism to describe it as presenting a **ratomorphic** conception of an individual's behaviour. Finally recognizing that people were not rats, environmental psychologists like Canter and Proshansky sought to distance themselves from this simplistic model of human behaviour to stress that people's actions can only be understood by appreciating how people's behaviour in place is mediated by cognitive processes which vary between individuals according to their social characteristics, intentions in place, and so on.

■ Exercise

Pile (1996, 19–44) offers a lucid but often critical account of the origins and diffusion of behavioural perspectives in geography, characterizing such approaches as 'robomorphic' (rather than ratomorphic) because of the way they ignored areas of the mind which seemed unknowable and unmappable, reducing human beings to the level of information-processing machines. On the basis of what you have read in this chapter, what areas of human existence and thought seem to be lacking in behavioural theories of environment and behaviour? Do you agree with Pile?

■ Summary

In this chapter, we have begun to explore how the relations between people and place are mediated by processes of information acquisition and processing that differ between individuals according to their age, bodily ability, gender, social role and so on. Important concepts outlined here include the following:

- Environmental perception refers to the acquisition of information from our surroundings via the senses of touch, taste, hearing, smell and sight. The use of these senses differs from person to person and also from place to place.

- Environmental cognition refers to the way this information is processed and organized by the brain. This process differs according to our reasons for being in particular places as well as our psychological development.
- Information acquired from the environment is built up into environmental images which are (relatively) stable and learned mental constructs which are used to orientate ourselves in familiar and less familiar places. Mental mapping techniques, distance-estimation exercises and questionnaires are important means of exploring these images.
- Behaviour in place is shaped by our perception of its physical characteristics, including the relationships between objects and people.
- Behavioural methods recognize that each individual organizes and understands their surroundings in a different way, but seeks to make meaningful generalizations about processes of cognition using research techniques which explore people's subjective view of the world, rather than the world as it 'really' is.

four A sense of place

This chapter covers:

4.1 Introduction

As we suggested in Chapter 2, it has become commonplace to suggest that we live in a globalizing world. This implies that all places and all people are increasingly tied in to global structures and institutions (such as transnational corporations or the United Nations), where we experience 'time–space compression' (the world appears to get smaller because travel and communication are faster and cheaper than in the past) and where communications via mobile telephones, satellites and the Internet can overcome the 'barriers' of distance which once restricted human interrelationships. While suggestions of globalization can be disputed on the grounds that they don't take into account those people who do not have access to intercontinental travel or computers (many people in so-called Less Developed Countries, for example, or deprived social groups in wealthy countries), it is certainly the case that 'big business' and communications technology have permeated many parts of the world, leading to the global extension of particular landscapes, economic systems (especially of monetary exchange) and social practices (in relation to musical and fashion tastes, for example). It could possibly be argued (but not in this book!) that 'everywhere is becoming the same' as a result of processes of globalization. As we discussed in Chapter 2, the implication here is that places, as individual and unique entities, may be wiped out and rendered meaningless in a globalized world where the same styles of building, economic activities, leisure activities and so on appear everywhere. The concept of the 'non-place' has arisen from the feeling that the authentically locally specific has been erased in favour of places like international airports which look very much the same everywhere (Augé, 1995). As Benko (1997, 23) suggests, 'Never before in the history of the world have non-places occupied so much space'. Given these ideas, what are we to make of the concept of 'place'?

At this point, we might also think back to Chapter 1 where we considered the idea of geography as spatial science. Inevitably, this appeared to make everywhere seem the same as the complexity and sheer messiness of the Earth's surface was reduced to an 'isotropic surface' in the quest to determine rules or laws of spatial behaviour. Places became simply nodes in networks of rationally determined flows of people, commodities and money. According to Stephen Daniels (1985, 144), this type of approach to studying the relationships between people and places is 'intellectually deficient', meaning that the reduction of places to nodes, and of people to rational economic men and women, misses out on a number of important dimensions of the relationships between people and place. Behavioural approaches to human geography seek to redress this by developing more multidimensional understandings of what it is to be human. However, as we indicated in Chapter 3, behavioural geography and environmental psychology can be accused of reducing people to decision-making machines, rather like robots, potentially missing out on the depth and quality of the relationships that exist between people and *their* places. It is telling that in the phrase 'people and their places' some idea of ownership and creativity is implied; here, we can start thinking about how people are involved in 'making' places in various ways that generate (and sustain) emotional attachments between people and places.

What we begin to discuss in the remainder of this chapter is a series of responses to the two sets of issues outlined above. These responses have been collectively referred to as constituting a **humanistic** approach to human geography which seeks to reconceptualize place in the context of human experiences of living in the world. In essence, this a perspective that emphasizes how the distinctly human traits of creativity and emotion are involved in the *making* of place. Drawing on such ideas, what we try to do in this chapter is to demonstrate that people's attachments to place are profoundly important (even in the context of a globalizing world), requiring us to explore people's individual relationships with the world to a far greater depth than is possible using quantitative techniques. You will notice that in this chapter, humanistic approaches are generally presented in a positive light. You should, of course, be immediately suspicious of this, and following this chapter, Chapter 5 extends our discussion to demonstrate some additional dimensions of this type of approach and explores some important criticisms of humanistic geography.

4.2 Regional geography, home places and humanistic approaches

Exercise

Before we go further think about the following questions . . .

Think about those places which are special to you personally; make a quick list. These could be at different scales: room, garden, house, village, seashore, mountain, city or country (for example). What is it that makes you feel that these places are uniquely special? What can you say about the *genius loci* of these places – i.e. what is their unique 'spirit of place'? What do *you* bring to these places in your encounters with them? How does your 'sense of place' affect the ways that you behave and the things that you do in these places?

We mentioned above that humanistic geography is at least partly a reaction to the quantification of geography during the 1960s. Part of that reaction was to look backwards in time, to earlier traditions of **regional geography** which emphasized careful description of the character of particular regions (see Chapter 1). For example, David Ley's (1981) humanistic writing drew on the work of French regional geographers such as Vidal de la Blache who were writing at the start of the twentieth century about the distinctive *genres de vie* (ways of life) which bound people to their homes. Importantly, this 'home' was regarded as the synthesis of the physical (the environmental characteristics of a region) and the social (the ways in which people lived and organized their lives). While Ley (1981, 221) was critical of the way that regional geographers seemed to promote environmental determinism, he found great value in the way that a home place was conceptualized as having 'both materialist and idealist attributes; it was both a thing and an idea'. What he stressed here was the inseparability of the 'real' (material) attributes of a place and its 'imagined' idealized attributes. Ley was thus strongly opposed to the **naturalist** assumption that social phenomena could be studied in the same way as physical phenomena – by looking for general laws or rules and explanations of cause and effect. Clearly, quantitative (positivist) geography, with its laws of spatial science, could be seen as adopting this naturalist perspective. The same criticisms were also extended to the behavioural geography discussed in the previous chapter – especially in relation to the ideas of 'rationality' underpinning decision-making and cognitive processing of environmental information.

From a humanistic perspective, naturalist approaches overlook anything that cannot be measured and treated in the same way as natural or physical data (treated to statistical analysis, for example). The great richness and diversity of human experience – the wealth of things it means to be human – thus becomes invisible. How, for example, do you measure the concept and experience of 'home'? How do you quantify what it feels like to sit on a sandy beach watching a beautiful sunset? Pierce Lewis (1985, 468), for example, described a 'love affair with those Michigan dunes . . . [which] had everything to do with violent immediate sensations: the smell of October wind sweeping in from Lake Michigan, sun-hot sand that turned deliciously cool when your foot sunk in, the sharp sting of sand blown hard against bare legs . . . One is meant to feel those landscapes, not analyse them'. For David Ley, and other humanistic geographers, the reinsertion of human experience, feeling and emotion into the very heart of human geography was an essential precondition for

a fuller (and ethical) understanding of the relations of people and place. As Daniels (1985, 145) puts it, 'Humanists reject the reduction of space and place to geometrical concepts of surface and point; humanistic conceptions of space and place are thick with human meanings and values.'

To expand on these ideas, it is necessary to go further into the theoretical frameworks underpinning humanistic geography. Hopefully this will enable us to show how geographers have taken ideas developed outside the discipline (principally in philosophy) and used them to interpret the relations between people and place. We begin, therefore, by exploring **existential** and **phenomenological** ideas. This pair of rather intimidating terms covers some difficult philosophical ideas about the relationships between human beings and the world in which they live. We will try to provide a brief summary of each, and then pull out their implications for humanistic approaches to geography.

Existentialism is an idea associated with, for example, the French philosopher and novelist Jean-Paul Sartre. It is a philosophy which stresses the specificity and uniqueness of each individual's experience of the world. Richard Peet (1998, 35) explains that 'existence for existentialists is characterised by concrete particularity and sheer "givenness", as compared with the abstract and universal concepts of humanity and life common to positivist thought'. From their perspective, abstract, 'high-level', theorization, which turns diverse landscapes into an isotropic surface and very different people into rational economic men, can be accused of offering a generalizing and naturalistic view from 'above'. Instead, existentialism demands a locally specific view from 'below'; a grounded view exploring the concrete and particular perspectives of individual people in specific places. This perspective is 'given' in everyday encounter with the world, and exists prior to any abstract conceptualization. The German philosopher Martin Heidegger (1890–1976) used the German word *dasein* to emphasize that what is important to human experience is **being in the world** (where 'being in' is opposed to reflecting upon). For existentialists, this is the key to understanding the relationship between people and the world. 'Being' is characterized by existing physically in the world – taking up physical space and existing *in relation to* other physical objects (including other people). From this perspective, it is the **relational** encounter with the world that brings the world into existence for each person. People's physical relation to things, therefore, affects the way that they organize and make sense of their worlds. This means that our knowledge of the world can, firstly, be said to be *created* by us (rather than something we simply discover) and secondly, results from our **encounters** with things (which are, for example, in front or behind, above or below, our bodies).

In essence then, existential ideas propose that humans create their worlds by making meaningful the physical phenomena – other people, places and objects – they encounter as they move through geographical space. For thinkers like Sartre, this projection of meaning on to the world was related to our sense of separation, estrangement or alienation from the world (described as an essential part of the human condition). This **existential dread** results from the feeling that we are completely different to everything else we experience, so that we attempt to make the world of objects comprehensible to ourselves by giving these objects meaning (Csikszentmihalyi and Rochberg-Halton, 1981). Objects, people and places thus become meaningful to us, while the systems of meaning that develop through this process become an essential part of the world as we experience it (something we return to in Chapter 7 when we consider the way objects can bear meaning). This is something you are no doubt aware of – all of us tend to invest everyday material objects with personal meanings (and possibly own lucky mascots or items of clothing that have special meaning for us).

Phenomenology is related to existentialism, and can be seen as a methodology (a way of studying) as well as an interpretative framework (a way of knowing). This is an approach that suggests the best way to find out about human relationships with the world is to use intensive forms of description. As with existentialism, individual human experience is central to this description. Phenomenologists reject the naturalist scientific assumption that an underlying 'reality' can be studied and described independently of human experience. Instead, they suggest that our experience is *itself* an essential part of reality, and there is no separate 'real' world external to human experience. In saying this, the founders of phenomenological thought, such as Edmund Husserl (1859–1938), sought to overcome the often assumed dualism between mind and matter (subject and object). This dualism, firstly, separated the human consciousness from a supposed 'real' world and, secondly, implied that this 'reality' could be studied independently of human experience. Instead, phenomenologists see the world in terms of phenomena that come into existence through human experience of them. This experience can be subconscious or unconscious, and is always associated with the human subject's **intentionality** towards objects – meaning a person's intention to use or interact with objects (which could be anything from a shovel to another human being; the flat blade of metal attached to a wooden pole only becomes a spade when you approach it with the intention of digging a hole). This intentionality happens spontaneously, all the time, as we live and exist in the world. Phenomenologists aim to recover the moment of intentionality in order to strip away the accumulated layers of conscious meaning and conceptualization (including academic theorization) that hide the 'true essences' of the initial moments of encounter with phenomena. Describing these essences involves using human sensory relations with the world (seeing, hearing, smelling, tasting, touching), but also mental relations such as remembering, imagining and having emotions. Husserl implicated all of these things in what he called the **lifeworld**. This is described by Peet (1998, 39) as the 'moving historical field of lived experience', implying that our experience is constantly changing (as we move through the world – although in this context we could also refer to being emotionally 'moved' by the world) as we live and do things in the world. We return to an explicitly geographical understanding of the lifeworld below (section 4.4) with a discussion of David Seamon's work, but it is worth mentioning here that place is understood as being a key part of the lifeworld of individual people (see also Buttimer, 1976).

The ideas discussed above are somewhat complex, yet at this stage we can summarize some of the key concepts which have emerged from phenomenology and existentialism to inform the development of humanistic geography:

- Humanistic thinking rejects naturalism. For humanists (and hence humanistic geographers) there is no world outside of human experience. The reduction of the world to the measurable and quantifiable is, as suggested earlier, theoretically suspect.
- Humanism stresses the individuality of human beings, in experiencing and creating their own worlds.
- Humanistic method focuses on understanding the intricacies of human experience and 'being in the world' in an attempt to get behind abstract theorization and uncover the 'true essence' of people's encounters with phenomena during their everyday lives.

■ Reading

Before we turn to show how these ideas have been developed by geographers in particular research contexts, we might advise that you begin to think through this yourself by looking at a paper informed by humanistic ideas and considering what makes it distinctive as a humanistic piece of research. We particularly suggest you look at the paper by Douglas Pocock (1996) entitled 'Place evocation: the Galilee Chapel in Durham Cathedral'. This is a descriptive piece written from an explicitly phenomenological perspective, demonstrating humanistic geographers' great concern with specific and highly meaningful *places* (often described as 'sacred' places). Pocock, himself immersed in this place, discusses how various people visiting the Galilee Chapel encounter and relate to a place which is saturated with historical and religious meaning.

■ 4.3 Humanistic geography: 'there's no place like home'

As we noted above, humanistic theory suggests a focus on place as opposed to abstract space. It has been the case, therefore, that when humanistic geographers look at the human environment, they have seen it as consisting of innumerable places within which people live, and to which they attach meaning. This attachment of meaning can be thought of as a way of bringing places into the ambit of human understanding – making a place meaningful makes it *belong* to us in some way. Simultaneously, meaningful places become part of who we are, the way we understand ourselves and, literally, our place in the world. In other words, our meaningful relationships play an important part in the formation of our **identities**.

Humanistic geographers writing in the 1970s were thus very much concerned with refocusing geography's attention on to place as a deep and complex part of our experience of the world. Suggesting that places as 'objects' can be examined in relation to our intentionality (how they are to be used or related to), they attempted to produce geographical knowledge which emphasized the ways in which places can have a great deal of meaning and significance for people. This new geographical knowledge hence contrasted with that generated via quantitative geography (by emphasizing place rather than abstract space) and also behavioural geography (by beginning to focus on unconscious phenomenological relationships with place rather than just on cognitively mediated perceptions of place).

This set of ideas can be taken a step further by considering how specific geographers developed humanistic concepts. Firstly, we will say something about how Yi-Fu Tuan conceptualized place and, secondly, we expand on Edward Relph's understanding of the 'authenticity' of place. From there, we will be in a better position for thinking about the importance of 'home' places for humanistic geographers. As we do this, however, you should take time to think about your own experience of 'being in the world'. How far does what humanistic geographers have to say fit in with your own experience? Are there any immediate problems you can identify?

We start here with the seminal work of Yi-Fu Tuan, a geographer whose writing has often blurred the distinctions between prose and poetry. A key feature of Tuan's humanistic geography is the way that social and geographical understandings of place are inseparable. Place is used by Tuan to refer not only to geographical location, but also to social position – as, for example, when somebody is told 'it is not your place to behave in that way'. As a quick (and simplified) example, think about the social and geographical relationships in the offices of a business. The boss may sit behind a large and imposing desk in a big office behind closed doors; to gain access to her, one has to report to a receptionist's desk outside in a lobby. The boss is in a position of relative social power or authority, while the receptionist is in a subordinate position, having to obey orders 'from above'. Both social and geographical place are important in this relationship (the boss's 'higher' social place, or **status**, is reflected by her private and imposing office, while the receptionist's 'lower' place is represented by his desk outside of the main office). Further, this socio-spatial relationship is significantly implicated in the way that the two people involved behave towards each other within the business context (something we considered in the context of a classroom in Chapter 3). This is most easily illustrated by thinking about the consequences of a disruption to 'normal' expected behaviour: if, for example, the boss was called in to take a letter by the receptionist, or if the receptionist started issuing commands about future company strategy. This would be behaviour that we might refer to as 'out of place', and again the word 'place' has both geographical and social connotations. You can probably refer to other spatial **metaphors** which combine geographical and social meanings in such a way as to make them inseparable. What do we mean, for example, when we say we feel 'close' to another person? Or when we feel 'distant'?

The case discussed above is really an example of place being important at the microscale. However, as Tuan (1974, 245) says: 'Places can be as small as the corner of a room or as large as the earth itself'. He goes on to suggest that we can think about and experience place in two ways (outlined below), but also that in the contemporary, globalizing world, to our loss, both of these ways have been disrupted. Part of his humanistic project is to reclaim the depth of experience of place that it is claimed we need to be fully human. Tuan's first way of thinking about place is to suggest that certain places have their own unique 'spirit and personality'. These are highly imageable places that, for example, command awe in the human observer – perhaps places which are sacred within different religious or mythological traditions, like cathedrals or Ayers Rock. These types of place might also be thought about as **public symbols**, experienced in similar ways by a specific group of people. For example, in the UK, places like the Houses of Parliament act as icons, symbolizing British national identity and a tradition of democracy. Places like Trafalgar Square, commemorating a notable British naval victory, publicly symbolize the idea that 'Britons never shall be slaves'. The Millennium Dome in London is a recently created public symbol, trying to create a sense of national unity through association with a special event (we return to a consideration of national identity in Chapter 5). At a larger scale, places as large as cities or countries can symbolize specific things to particular groups of people; consider, for example, how Oxford and Cambridge are public symbols of 'learning' in the UK, or the way that Cuba came to symbolize a communist 'threat' to the USA in the 1950s/1960s.

Clearly, public symbols rely on stereotype and imagination – not everyone living in Oxford is an academic (there are areas of industrial manufacturing and social deprivation, for instance) – but as easily recognized **icons** they are important beyond the 'reality' of what or who exists in these places. Tuan says that these places give up, or project, their meaning to the eye (see Tuan, 1977). This visual encounter between human and place is enough to demonstrate something to the human, provided that that human is part of the specific group of people to whom the symbol is directed (that is, a group with a shared awareness of the meaningfulness of particular icons). Tuan, in defining this type of place, was especially concerned with sacred places, discussing the ways that in the ancient world landscapes were rich in places which held sacred, religious or mythological significance. It must be borne in mind that what is sacred to one group of people may not be seen in the same way by other groups, as the following quotation from Amos Rapoport demonstrates:

> Many Europeans have spoken of the uniformity and featurelessness of the Australian landscape. The aborigines, however, see the landscape in a totally different way. Every feature of the landscape is known and has meaning – they then perceive differences which the European cannot see. These differences may be in terms of detail or in terms of a magical and invisible landscape, the symbolic landscape being even more varied than the perceived physical space. As one example, every individual feature of Ayers Rock is linked to a significant myth

> and the mythological beings who created it. Every tree, every stain, hole and fissure has meaning. Thus what to a European is an empty land may be full of noticeable differences to the aborigines and hence rich and complex.
>
> (Rapoport, 1972, 3–3–4)

In the contemporary, arguably more secular, Western world the sacred meaning of sacred places is often diminished. Nevertheless, traces of such meaning do linger in sacred places – think about, for example, the way in which even non-believers can feel a sense of awe in a cathedral or church, to the extent that their behaviour is modified (i.e. they move slowly and talk quietly). Here, we could also mention Stonehenge (see Figure 4.1), a site whose pagan/druidic significance has, to most, been lost, but which remains a site of touristic pilgrimage.

Tuan's second way of conceptualizing place moves us from the public to the personal, through the concept of a 'sense of place'. Instead of places having symbolic meaning 'within' themselves (which can be imparted to any observer who enters into an intentional phenomenological relationship with them), a sense of place relies far more on the individual, with places becoming significant places for them alone. To develop a sense of place requires that one knows the place intimately and reacts to it emotionally (rather than rationally). This is, perhaps, what Pocock (1996, 379) means when he states that 'an epistemology of the heart concerns knowledge acquired by union or communion' in his paper on the Galilee Chapel mentioned

Figure 4.1 Sacred space and symbolic place? Stonehenge

Photo: C. Rowley

above (an epistemology is a framework for thinking about how we construct knowledge – see Chapter 10). The implication here is that intimate knowledge is gained over a long period of time through an extended encounter with place. In turn, this invests the individual with a deep sense of that place, making the place an extension of the individual. In contrast to public symbolic places, the places which are meaningful for specific individuals are very difficult to identify, because there are not necessarily any clear outward indications inscribed in the place itself. Instead, for the individual, these places are known and cared for from within, existing as what Tuan calls **fields of care**. Further, Tuan coined the term **topophilia** to express the phenomenological encounter between individual and field of care (the word 'topophilia' is the result of combining two Greek words to mean 'love of place'). This implies that individuals have an emotional need to identify with often personal and intimate places, and hence 'construct' these places for themselves on the basis of repeated experiences (of the sounds, smells, sights and sensations encountered in a place), the formation of behavioural routines (such as a journey to work or college, or a familiar route for walking the dog around the local area) and ties of spirituality and kinship (involving, for example, religious belief and family connection).

A sense of physically being and feeling 'in place' or 'at home' can then be regarded as a sign that an individual has established an emotional tie to a place. This emotional bond is necessary for cultivating a sense of place, along with the requirement that an individual needs to be physically intimate with the place at the same time (i.e. physical engagement with the place doesn't have to be thought about, but is embedded in the body of the individual). But, as with his discussion of publicly symbolic places, Tuan suggests that this type of emotional and physical connection with place has become less apparent in contemporary society. In his thesis, globalization has produced a relatively bland environment where the opportunity to find a place of one's own is severely curtailed. He therefore notes a progressive decline in people's sense of place in contemporary Western society, linked perhaps to a propensity to travel quickly from one place to another in the isolated, detached comfort of cars, rather than walking and loitering in a way that encourages the formation of distinctive senses of place. Tuan (1974, 245) stresses this when he argues that 'places are locations in which people have long memories, reaching back beyond . . . their own individual childhoods to the common lores of bygone generations . . . . Time is needed to create place.' There is, perhaps, a sense that Tuan is looking back nostalgically to a 'golden age' when people were more settled and before their lifeworlds had been colonized by (over-)structured economic, political and social processes. Here, modern planning and redevelopment was proposed by Tuan to be a major force obliterating distinctive place identity. We will return to such ideas later (suggesting that local distinctiveness and people's attachment to place are still of fundamental importance in the contemporary world).

In *Place and Placelessness* (1976), Edward Relph's line of argument follows Tuan in suggesting that, firstly, a sense of place is important for individual (and also community) identity, and secondly, that this sense of place has been lost or degraded in the modern world. It is as a result of this loss or degradation that Relph suggests that the modern geographical experience is one of **placelessness**. Despite this, Relph (1976, 1) maintains that 'To be human is to live in a world that is filled with significant places: to be human is to have and to know *your* place', and, as with Tuan, place is understood as both social and spatial. Relph extends his discussion by thinking about the ways that a sense of place can be either authentic/genuine or inauthentic/artificial. 'Authentic' sense of place implies that a fundamental, lasting truth about a place is known, going beyond the ephemerality of the constantly changing modern world, and tapping in to an unchanging *genius loci*, or unique 'spirit of place'. In Relph's own words, having this authentic sense of place implies 'a direct and genuine experience of the entire complex of the identity of place – not mediated and distorted through a series of quite arbitrary social and intellectual fashions about how that experience should be, nor following stereotypical conventions. It comes from a . . . profound and unselfconscious identity with place' (Relph, 1976, 64). The alternative here is a superficial knowledge of (and relationship with) place, and a landscape which is a bland 'flatscape' rather than one full of significant places.

These ideas are closely tied up with the concept of **home**, which is central to humanistic geography. To have an authentic sense of place (at a range of scales, from a house, to a region or country, or even the whole Earth) is to have a sense of **belonging** – you talk about your home town or home region, for example, and this implies a very deep sense of attachment, making place a strong part of who you are and the way you think about yourself. As Lupton (1998) contends, the home is often idealized as a 'territory of the self' that contrasts sharply with the chaos of the (outside) world.

■ Exercise

Which places would you call 'home'? Why? Think about this at different geographical scales – the house, city or country, for instance. If you have moved places to go to university or college, would you call the place you came from or the place where you are studying 'home'? What makes a home – what are its essential features for you personally? When you say you feel 'at home' in a place, or when you tell a guest to make herself or himself 'at home', what *exactly* do you mean by this? What we might suggest is that being 'at home' in a place is less to do with a simple physical location, and much more to do with deep and meaningful relationships between a person and the place which have developed over time.

You may conclude that although home places seem to be those ordinary, everyday places, where for most of the time you carry out relatively mundane everyday activities, they are nevertheless places which are saturated with meaning.

The cultural significance of 'home' in contemporary Western society is indicated, using examples from the UK, by the rising number of home improvement programmes (such as 'Changing Rooms' or 'Home Front'), magazines (such as *Ideal Home* and *House Beautiful*) and shops like Habitat and Ikea which sell the materials (and flatpack furniture) needed to emulate the styles and lifestyles promoted by these magazines and television programmes. A home becomes something to be invested in, serving as it does as a locus for the identities of those who live there. Similar investments of time, money, emotion and labour can be seen in the private gardens attached to (and very much part of) the home. As with the house itself, there is a plethora of magazines, television programmes and retailers dedicated to providing 'new ideas' designed to transform gardens into 'outdoor rooms' (a common motif used in the programme 'Ground Force', where more emphasis is placed on garden furniture, lighting and surfacing than planting). Cynically, we might note that such programmes and their resultant designs are very much driven by commercial trends, mediated by the various interior decorators and landscape gardeners employed on the programmes rather than being the result of the occupants' imaginations and creativity. Nonetheless, the time and money which people are prepared to invest in personalizing their homes does suggest that home spaces are considered more than just 'a roof over one's head' (see Putnam, 1993).

Another key idea that might arise in association with the concept of home is a sense of geographical segregation or separation. At different scales, 'home' is often understood as a place within which only certain people and things belong; it is a place to which a person or group of people can withdraw from the outside world, as if into a castle. As Bachelard (1958) puts it, the house creates 'a private space for dreams'. But an additional dimension to this is that a sense of geographical segregation from the outside world often implies a sense of **social segregation**. For example, a house or flat where a person lives is made into 'home' partly through their ability to spatially exclude certain people (who maybe don't belong to his or her family). Here, the consequence of being 'at home' translates into the idea that the individual can feel in control, free from the intrusion of others. This argument can, perhaps, be extended with reference to things like national boundaries – think, for example, about how nations monitor their boundaries by restricting residence to certain groups. An important way of thinking about these things is to consider identity in terms of the concepts of 'self' and 'other'. Your identity – the way you think about yourself and the way others think about you – is defined not just by what you are (for example, in terms of your gender, sexuality, skin colour or social class) but also what you are not. There is a consequent tendency for people to try to exclude those who are 'other' to themselves from their home places. Here, you might start to discern some nasty cracks appearing in the rather idyllic version of home we've begun to sketch out (something we will come back to in the next chapter).

Reading

Yi-Fu Tuan's writing is very readable, if esoteric. 'Space and place: a humanistic perspective' (1974) and *Space and Place: The Perspective of Experience* (1977) are good insights into his humanistic geographies. Lily Kong's article 'Geography and religion' (1990) provides greater detail on sacred spaces and their role in the contemporary world – a somewhat neglected tradition in geographical research.

4.4 The geography of the lifeworld

The title of this subsection is taken directly from a book of the same name by David Seamon (1979). Following Tuan and Relph, Seamon insisted that a satisfactory human existence requires the individual to have strong links to place and locality, but that globalization processes had reduced the potential for this to occur. Seamon was thus concerned with how individuals' phenomenological immersion into a geography of everyday life constructed their lifeworlds (a term we defined in section 4.2). To explore people's lifeworlds he argued that it was necessary to consider the relationships between people's behaviour (what people do in places) and their experiences of place. Seamon (1979, 16) defined the methodology of phenomenology as 'a way of study which works to uncover and describe things and experiences – i.e. *phenomena* – as they are in their own terms', rejecting abstract theorization and categorization (e.g. making spatial laws or categorizing people by gender) in favour of an holistic approach receptive to all forms of human experience, emotion and behaviour. On the basis of this, he developed a framework for understanding the relationships between people and place, suggesting that three interlinked themes are significant:

- **Movement:** a focus on how the individual (and especially her or his body) move through space, on a day-to-day routine basis (something we began to think about in Chapter 2).
- **Rest:** a focus on human attachment to place – clearly the concept of home is highly significant here.
- **Encounter:** a focus on the ways in which people notice and observe their everyday worlds.

Figure 4.2 illustrates how these themes can be related in a 'triad of environmental experience'. What we need to stress here is the particular focus on the **body** as something which can 'know' its environment on its own terms. This may be an uncomfortable idea because we are so used to thinking about the ways in which our minds are our centres of experience, feeling and knowledge, but Seamon suggests that our

Figure 4.2 Triad of environmental experience

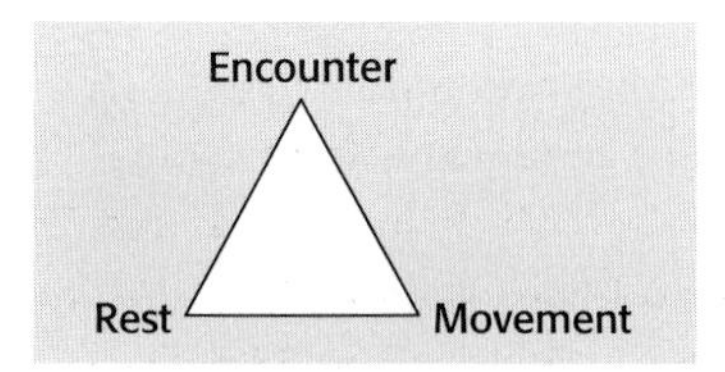

Source: Seamon (1979)

bodies too have intimate knowledge of the everyday spaces of our lives, so that we do not need to continuously think consciously about how we interact physically with the world. There is a dualism in much of Western thought which separates the mind from the body – the body is seen as largely under the control of the mind, as the tool of the mind (see Butler, 1999). Seamon is keen to break down this dualism. Instead of the human subject being theorized as a mind 'trapped' in, or working with, the body, he suggests that we consider individual people as 'body-subjects'. He employs an interesting descriptive metaphor for thinking about how individuals use space in their everyday routines – that of the **body ballet**. This suggests that (to a certain extent) routine and repeated movements in familiar places can be thought of as choreographed (i.e. performed without thinking, as opposed to movements being constantly and consciously thought out in the mind, before being passed on as instructions to the body). Extending this metaphor to cover groups of people interacting with each other in particular places, Seamon uses the phrase **place ballet** to describe the choreographed but complex movements of several bodies simultaneously; something which can be observed in everyday street scenes (Figure 4.3). Unlike behavioural geography, where movement is reduced to something which can be measured and perhaps explained with reference to cognitive process, for humanistic geographers movement becomes something to be considered as a phenomenon in its own right – as an essential and dynamic part of human experience (Buttimer, 1976).

So what does all this mean in terms of our relationships with place? Seamon, using the results of a large amount of intensive field work where he discussed everyday experiences of space and place with focus groups, draws some important conclusions from his movement–rest–encounter framework. Here, we want to emphasize three main points. Firstly, Seamon suggests that people are attached to place both emotionally and bodily; this multidimensional attachment is a key part of being 'at home' in a place. Secondly, he proposes that attachment and 'at homeness' tend to be associated with routine, regularity and the everyday. These notions become taken for granted by both body and mind so that conscious thought appears unnecessary in our everyday dealings with familiar places and/or situations. A sense of continuity is important here, and many people express strong urges to preserve what they know – despite recognizing that places must undergo substantial changes

Figure 4.3 Street ballet

Photo: © Tony Stone Images

(Lynch, 1972). But even unfamiliar places can be dealt with in the same way by the body-subject, especially when they remind us of places that are significant to us. For example, you would know almost instinctively how to behave as a guest in somebody else's house even if you had never been there before, because you know how you would wish them to behave in yours. You might even feel 'at home' in an unfamiliar branch of a familiar supermarket, because the layout, the colour scheme, the products and the ambience are similar to the branch you normally use. Thirdly, Seamon asserts that people encounter the world – come into contact and engage with it – by moving and resting in it (Figure 4.4 shows this in diagrammatic form). He suggests that there is an important dialectical relationship between movement and rest ('dialectical' here simply means considering the two as opposites of each other, but at the same time bound together in a relationship of tension). The security and 'centredness' of home is contrasted with the sense of adventure and perhaps danger of venturing into the unknown. As a result, it is important to recognize that within each individual there are simultaneous desires to be both safely 'at home' and to escape the confines of home and move into the unknown. Yet, as we have seen,

Figure 4.4 The dialectic between movement and rest

Source: Seamon (1979)

notions of home are not fixed, so that movement away from home can eventually result in an expansion of what is considered as 'home'; the 'unknown', in effect, becomes familiar.

■ Reading

Seamon's book, *A Geography of the Lifeworld* (1979), examines his concepts of movement, rest and encounter in some detail, providing evidence from his own research into people's use and experience of space. Writers such as Jane Jacobs have developed similar arguments in studies exploring the choreographed rhythms of the streets; see *The Life and Death of Great American Cities* (1961). On a more extensive scale, Teather (1999) discusses the way bodies move (and are moved) across the life-span.

■ 4.5 Writing home: place, landscape and belonging

Having discussed the nature of humanistic geography in some depth, we wish now to think about the ways in which some of these ideas have been applied in different contexts. We begin by considering the way in which literature has been used by humanistic geographers to examine the emotional and bodily relationships that exist between people and place. This focus on literature is not particularly surprising given the variety of ways in which people have expressed their feelings about place through creative writing (as well as music and painting). Indeed, creative writing has enabled many authors to express something about the *genius loci* of places which are special to them for a range of reasons. The intention has often been to evoke in the reader a sense of place about a place they have not or cannot ever physically go to (where, for example, the place is fictional, or set in a historical context). Such writing is often intensely personal and deeply descriptive. The excerpt from Laurie Lee's famous *Cider with Rosie* (first published in 1959) below is, perhaps, a good example of the creative evocation of place. As you read it, think about the way Lee uses language to create a multidimensional rendering of place as seen, heard, smelt, touched and felt.

> The village to which our family had come was a scattering of some twenty to thirty houses down the south-east slope of a valley. The valley was narrow, steep and almost entirely cut off; it was also a funnel for winds, a channel for floods and a jungly, bird-crammed, insect-hopping sun-trap whenever there happened to be any sun. It was not high and open like the [River] Windrush country, but had secret origins, having been gouged out from the escarpment. . . . Like an island, it was possessed of curious survivals – rare orchids and Roman snails; and there were chemical qualities in the limestone-springs which gave the women pre-Raphaelite goitres [an enlargement of the thyroid gland]. The sides of the valley were rich in pasture and the crests heavily covered in beechwoods.
>
> Living down there was like living in a beanpod; one could see nothing but the bed one lay in. The horizon of woods was the limit of our world. For weeks on end the trees moved in the wind with a dry roaring that seemed a natural utterance of the landscape. In the winter they ringed us with frozen spikes, and in summer they oozed over the lips of the hills like layers of thick green lava . . . .
>
> I remember, too, the light on the slopes, long shadows in tufts and hollows, with cattle, brilliant as painted china, treading their echoing shapes. Bees blew like cake crumbs through the golden air; white butterflies like sugared wafers, and when it wasn't raining a diamond dust took over which veiled and yet magnified all things . . . .
>
> [Before school] wild boys and girls from miles around – from the outlying farms and half-hidden hovels way up at the ends of the valley – swept down to add to our numbers, bringing with them strange oaths and odours, quaint garments and curious pies.
>
> (Laurie Lee, *Cider with Rosie*, 1959)

What can we say about this excerpt from a humanistic geographical perspective? Firstly, we might consider the way in which a sense of place, a *genius loci*, is built up in great detail. A sense of the valley as an isolated, self-contained and bounded place gives a feeling of great intimacy to the description. The village is described as both 'cut off' and 'like an island' – it feels withdrawn from a wider world (and the modern world), contrasted with the open country beyond. The village immediately becomes a very special 'home' place for the author, who tries to convey the rich experience of being in that place. Metaphorical language is used to add depth to this isolated sense of place, so that being there is like living in a 'beanpod', for example, suggesting security. Secondly, a multisensual experience is evoked; this place is not just seen, but sensed through sounds, smells, tastes and feelings on the skin. But we need to go beyond that – this place is sensed deeply and spiritually, and this is beautifully demonstrated through the description of movements, colours and landscape. Think about how the lighting is evoked, or how animals and insects are described. The landscape is literally brought to life as, for example, the wind is described as a 'natural utterance' – the landscape speaks to the writer. Thirdly, we can examine how the place becomes part of the writer's identity. This is particularly evident where the writer's identity (his self) is contrasted with what is outside of the village (what is 'other' to himself). The 'other' here (the 'wild' boys and girls, for instance) is geographically and socially separate, being portrayed as distant, strange and perhaps rather threatening.

What we must also bear in mind here is that this is a *remembered* childhood, and clearly presented in rather idyllic terms (as is the rest of the book, despite the occurrence of some rather shadowy events acting as precursors to an increasing loss of innocence). Childhood itself is portrayed as a period of innocence – protected from the outside, adult and dangerous world in the same way that the village in its isolated valley is geographically separated from the wider world. The point is that the spirit of place evoked here is in many ways a fiction (despite the autobiographical intent of the book), and that descriptive writing cannot ever be a clear, transparent and 'true' reflection of a reality. Instead, it offers a creative **representation** of a particular place and time, based on experience, imagination and memory. We return to discuss ideas about representation in detail in Chapter 7, but we need to think about the implications of this for humanistic geography's suggestion that it can uncover and communicate true *genius loci* (see Chapter 5).

■ Exercise

Below are some verses from a poem by the Welsh poet Dylan Thomas (1914–53). The poem 'Fern Hill' recollects the poet's boyhood experiences of a Welsh farm (stanzas 1, 2, 3 and 6 – the last stanza – are reproduced here). The full poem appears in the collection *Deaths and Entrances*, first published in 1946. Take some time to read and 'feel' (and enjoy) the imaginative and flowing use of language in these verses. Then, think about how a sense of place is evoked. In what particular ways is this place special to the poet? Be critical: how far can this be said to be a 'true' portrait of the farmyard? You might want to look at Tuan's (1978) 'Literature and geography: implications for geographical research', in Ley and Samuels (eds) *Humanistic Geography*, to help think this through. Additionally, Spooner's brief paper (1992) begins to look at the relationship between Philip Larkin's poetry and the city of Kingston-upon-Hull.

'Fern Hill', Dylan Thomas (1946)

Now as I was young and easy under the apple boughs
About the lilting house and happy as the grass was green,
The night above the dingle starry,
Time let me hail and climb
Golden in the heydays of his eyes,
And honoured among wagons I was prince of the apple towns

And once below a time I lordly had the trees and the leaves
Trail with daisies and barley
Down the rivers of the windfall light.

And as I was green and carefree, famous among the barns
About the happy yard and singing as the farm was home,
In the sun that is young once only,
Time let me play and be
Golden in the mercy of his means,
And green and golden I was huntsman and herdsman, the calves
Sang to my horn, the foxes on the hills barked clear and cold,
And the sabbath rang slowly
In the pebbles of the holy streams

All the sun long it was running, it was lovely, the hay
Fields high as the house, the tunes from the chimneys, it was air
And playing, lovely and watery
And fire green as grass.
And nightly under the simple stars
As I rode to sleep the owls were bearing the farm away,
All the moon long I heard, blessed among stables, the nightjars
Flying with the ricks, and the horses
Flashing into the dark.

[. . .]

Nothing I cared, in the lamb white days, that time would take me
Up to the swallow thronged loft by the shadow of my hand,
In the moon that is always rising,
Nor that riding to sleep
I should hear him fly with the high fields
And wake to the farm forever fled from the childless land.
Oh as I was young and easy in the mercy of his means,
Time held me green and dying
Though I sang in my chains like the sea.

As long as we apply some caveats, examining written descriptions of specific **landscapes** seems to offer valuable 'humanistic' insights into the relationships between people and place (see also Women and Geography Study Group, 1997, 167–99). Alongside place, landscape has become a key concept for humanistic geographers. It implies a sense of expansiveness and openness particularly related to the visual (although we can also talk about 'soundscapes' or 'smellscapes' – see Chapter 8). A landscape, then, needs somebody there to look at it, and, according to Denis

Cosgrove (1989, 121), involves a distinctive mode of looking, 'a way of composing and harmonizing the external world into a scene, a visual unity'. Landscape is, nevertheless, a contested concept, meaning different things in different contexts. We can, perhaps, simplify this into three aspects of landscape. Firstly, we can think about landscape as the observable surface of the Earth, and it is perhaps this mode of landscape which has most appealed to humanistic geographers. Secondly, we can see landscape in a more critical fashion as the visual expression of relationships between human society and the environment. For example, think about the differences between an agricultural and an industrial landscape and what this can tell you about the relationships between people and natural resources, as well as relationships between different groups of people. Thirdly, geographers have also been interested in the ways that landscapes have been represented or created in a number of ways, including on canvas, in words, and physically (through, for example, architecture and landscape gardening). Here we will say something quite briefly about a humanistic interpretation of landscape as a 'mode of seeing' and come back to the second and third ways of thinking about landscape in Chapters 5 and 7, respectively.

A key exponent of a humanistic examination of landscape is Edward Relph, a geographer mentioned already in relation to humanistic understanding of place. Relph (1987) defines landscape in simple terms as a context for our everyday lives, and brings a phenomenological perspective to bear on the description of landscape. He explicitly focuses on description of landscape, as opposed to explanation, saying that looking for the underlying processes and relations resulting in particular landscape forms is inappropriate because 'trying to explain landscape in this way is no better than a biologist killing an animal to find out what makes it live' (Relph, 1987, 153). We should, according to Relph, focus on the surface appearance of landscape, which is its primary feature. Again, we should perhaps be critical about this position, and return to other ways of dealing with landscape in subsequent chapters. Relph suggests that we can use a three-stage phenomenological method to study landscape:

- **Seeing**: a process of careful and observant looking or gazing, rather than a casual glance. Phenomenologically, this implies a visual encounter with a landscape.

- **Thinking**: devoting careful thought to the encounter. Relph says that this thinking should be 'responsive' and 'meditative', and draw on the observer's experiences and feelings about the landscape. One should respond to the essence of what it is one is looking at. The alternative, dismissed by Relph, is to think 'aggressively', which implies imposing preconceived theoretical and technical structures, and making judgements on the basis of those. Relph is proposing that we rid our heads of preconceived ideas, and experience the landscape for what it is. There are clear links here to what we said about phenomenology and existentialism in section 4.3.

- **Describing:** writing about the encounter and the thoughts which it engendered. Relph suggests that this description should be straightforward, clear and simple, rather than cluttered with unnecessary detail.

Using this type of methodology, it should be possible to engage with a 'spirit' of landscape at a very personal level, to describe what it feels like to 'be' in a particular landscape, whether it is an aesthetically pleasing rural or wilderness landscape, an urban industrial landscape, or even a micro-landscape (e.g. the interior of a shopping mall). Clearly, academic writing isn't the only way in which one's experience of landscape can be described. Many poetic and painted descriptions have become very well known as evoking landscape spirit, communicating something about what the poet or painter experienced in their encounter with landscape.

■ Exercise

Figure 4.5 shows a painting by Vincent Van Gogh. What can you say about the way that Van Gogh saw, thought about and described this landscape?

In the next chapter we will carry forward a lot of the ideas explored here, considering humanistic geography in a critical light and exploring notions of home, landscape and national identity from different perspectives. Humanistic geographers have rethought the way in which human geography is studied in immensely valuable (and often very attractive) ways – especially through their engagement with our individual encounters with place and landscape and their explorations of geographical

Figure 4.5 *Crows in the wheatfields*, Vincent van Gogh (1890)

Photo: Amsterdam, Van Gogh Museum (Vincent van Gogh Foundation)

lifeworlds. However, there are many other important things to say about place, and about the relationships between people and places, which add further dimensions to what we have said so far in this book, and we will move on to explore them in the chapters that follow.

■ Reading

Some of the readings suggested here touch on issues which we will expand on in Chapters 5 and 7. Barnes and Gregory (1997) 'Place and landscape' in *Reading Human Geography: The Poetics and Politics of Inquiry* gives a good introduction to the key ideas of place and landscape in human geography. Relph's (1989) chapter, 'Responsive methods; geographical information and the study of landscapes', in Kobayashi and Mackenzie (eds) *Remaking Human Geography*, deals principally with the phenomenological approach to landscape outlined above. Denis Cosgrove's writings on landscape have proved influential in the discipline: Cosgrove (1989) and Cosgrove (1985) are both useful reading.

Summary

This chapter has covered a lot of ground, moving from an emphasis on the changing importance of place in a 'globalizing' world, through existential and phenomenological theory to a detailed consideration of the concepts of place and landscape in humanistic geography. The chapter can be summarized in the following points:

- Humanistic geographers have been largely responsible for opening up new dimensions of geographical enquiry which focus on subjective human experiences of, and encounters with, the world. This has often been associated with **existential** and **phenomenological** philosophical approaches, emphasizing the importance of 'being in the world'.
- The concept of **place** is central to humanistic understandings of geography: human attachment to 'home' is of great significance to the relationships between people and 'their' places. 'Place' is to be thought of as referring simultaneously to geographical location and social status – it is a socio-spatial concept, also implying a relationship with what are considered to be 'appropriate' or 'inappropriate' behaviours.
- Our geographical **lifeworlds** consist of dialectically related rest and movement (home and security as opposed to the unfamiliar and insecurity), and continual, embodied encounter with the world. We can understand ourselves as 'body-subjects', as our bodies 'know' their

environments (and 'find their place' in them) at the same time as we mentally know and imagine them.

- **Landscape** is also a significant concept in humanistic geography, with our visual experience of landscape an essential part of being in the world. Creative writing and landscape-painting can be used to communicate the author's or painter's experiences of place in the creation of a sense of *genius loci*.

five Disturbing place

This chapter covers:

5.1 Introduction

In the previous chapter, a broadly humanistic set of approaches to thinking about the relationships between people and 'their' places was explored. It was suggested that people develop a deeply felt attachment to places at different scales, with 'home' places being particularly important in their experiences of the world. Simultaneously, it was stressed that experiences of place and landscape are multisensory, embodied and emotional, and that we can perhaps begin to understand something of these in the ways that people describe place through media such as poetry and painting. Excavating these, we often find strong sentiments of topophilia.

In this chapter, however, we want to suggest that, far from experiences of home always being positive and fulfilling, the idea of home (and the associated concept of **territoriality**) can have alternative and more disturbing associations. What we wish to do, then, is to start to think about how home places, territories and landscapes, can carry contrasting sets of meanings, which may imply, for example, that while a place or a landscape can be experienced positively by some people, for other people the same place might well be experienced in far less positive (and sometimes acutely negative) ways. Our discussion will begin by reconsidering the home, but will later leave this 'private', domesticated space to consider people's experience of other, perhaps more 'public', places. Simultaneously, elements of a **feminist** critique will be introduced here to suggest that humanistic understandings of place have often been structured and written from a masculine perspective. Similarly, we will consider the ways in which particular groups – such as gay people, young people or people of colour – may have been excluded from dominant understandings or imaginings of home and landscape. There are strong links between this chapter and ideas about the representation and contestation of place meaning (see Chapters 7 and 9). At the same time the chapter can also be considered as a critical response to the overview of humanistic ideas we presented in Chapter 4. As part of this, explicit criticism of some of the ideas mentioned in Chapter 4 will be made towards the end of this chapter.

5.2 Home sweet home . . .?

In Chapter 4, we discussed the way in which home places, at a range of scales, have been interpreted in very positive ways by some humanistic geographers. For David Seamon (1979), for example, the home was strongly associated with being at rest in a place which is known intimately. Accordingly, the home may be seen as part of a person's field of care, with the **domestic** home being acknowledged as the place where an individual (and often her or his family) is said to live. Frequently, the home has been regarded as a place to withdraw to, a place of rest and a place where one has a large degree of control over what happens. It is seen as the place where individuals can assert their identities, and 'be' what they want to be. Home places are also imagined as spatially segregated from the other spaces in which the individuals conduct their everyday lives – most typically this involves a separation of home from spaces of work. Finally, and significantly, the home is frequently regarded as a place of **privacy**, where people are able to be and do what they want, free from outside interference. The privacy of the home is almost sacrosanct in contemporary Western societies, and thus opposed to the public nature of places and spaces outside the home – e.g. places of paid employment, leisure or education.

This very privacy from the prying eyes of the general public and the institutions of the (modern) state has, however, permitted a range of more negative experiences of home to pass often unnoticed, obscuring the tensions which are undoubtedly experienced in almost all domestic places at one time or another. It may be recognized, for example, that while on the one hand, the home is desired as a place of safety, on the other hand it can also be experienced as constraining, a place from which escape is desirable (saying, for example, 'I need some space'). Perhaps, then, we need to be able to consider home places in rather more complex ways than was suggested in Chapter 4. One way in which we can do this is to consider how domestic work, sexual violence and unequal power relations within the home can subvert the idea of home as a place of relaxation, retreat and safety. However, we need to remember that between these positive and negative interpretations of home, there are more ambiguous interpretations of the nature of home and the role it plays in shaping human identities.

A good starting-point here is to follow many feminist geographers by considering the role of women in the home (e.g. Gregson and Lowe, 1995; Domosh, 1998).

A key observation here is that the separation of home from work cannot apply when the work of one member of the household occurs *in* the home place. In the UK and North America at least, **domestic work** (housework and childcare, for example) within the heterosexual family has often been a role assigned to women. Following Relph's suggestion from the previous chapter that 'to be human is to have and to know your place', we can perhaps consider this less as a benign statement related to being comfortable in a place one knows well, and more as a way of consigning particular people to particular places. Accordingly, the idea that 'a woman's place is in the home' can be conceptualized as a statement of masculine oppression rather than a recognition of a supposedly natural order. Clearly, changes in society have meant that more women leave the home to go out to work, and more men do a share of the housework, but even in households where both partners share the housework it is likely that the woman will do most of it. Where the home is experienced as a place to rear children, cook meals, do the ironing and minister to 'the head of the household' when he (*sic*) comes in from a hard day at the office, it is difficult to also experience it as a place of rest and retreat. We can start to see here how **relations of power** are as important in domestic places as elsewhere; in Western society men have often been able to constrain the spatiality and roles of women. A related theme here is the way in which 'non-conventional' domestic arrangements are imagined to contravene the family-oriented nature of a home, where the family implies a (married) man and woman, with their children. Such a model has led to social and political marginalization of, for example, single-parent families or gay/lesbian couples (Watson, 1986).

Other severely negative experiences of home can be used to extend these ideas. For example, child abuse, whether sexual or involving physical violence, has often been hidden by the myth of the home as secure retreat. This myth has led to situations where parents have become increasingly concerned about 'stranger danger' and the threat to their children on the streets (see Valentine, 1996), despite the fact that most cases of child abuse occur within the home, by close relatives of the child(ren) concerned. Domestic abuse between adults has also been hidden behind the myth of the marital home as private haven – in fact it has only recently been recognized in UK law that rape (non-consensual sexual intercourse) can occur within marriage. Previous assumptions that marital rape simply could not occur implied, in effect, the absolute possession of one person (usually a woman) by another (usually a man). Similarly, it is only relatively recently that violence within the home has been taken seriously by the police and judiciary in the UK. As 'domestic' violence, it has frequently been dismissed as a private matter, occurring within the privacy of the home, in which a state institution – the police – had no right or cause to interfere.

Such instances of domestic abuse should lead us to rethink the understanding of home places gained from Chapter 4. For many, home places are experienced negatively, with the general interpretation of them as positive and protective sites obscuring realities of domestic abuse and subjugation. However, the idea of home cannot simply be understood in terms of a polarity of 'good' and 'bad', as home spaces are inevitably subdivided and compartmentalized between different household members. This compartmentalization often indicates something about the relationships between these household members, giving us clues about the way that experiences of home vary for different social groups. A key concept here is that individuals have a desire to **purify** their own space, making it 'off-limits' to certain people (or animals, plants, objects . . .) at certain times. In his book *Geographies of Exclusion*, David Sibley (1995) presents us with the idea that the home provides a key example of how the human desire for a purified environment results in rituals centring on the expulsion of 'threatening' materials. Accordingly, the task of 'housework' basically involves a process of removing apparently threatening and waste materials – such as dust, mites and 'dirt' – from the house; similarly, certain areas of the house (e.g. the sink, the waste bin, the toilet) are reserved for the expulsion of material and faecal wastes (Figure 5.1).

For Sibley, these purifying desires are bound up with a series of crucial binary oppositions in Western society between clean and dirty, tidy and untidy, inside and outside, public and private. Sibley's analysis of the home as a purified environment stems from some aspects of **psychoanalysis** which suggest that the behaviour and experience of the individual derives partly from an **unconscious** part of the mind. As soon as we start to talk about 'the unconscious' here, we need to begin to rethink somewhat what we mean by the individual human subject. Much of our earlier conceptualization of what it means to be human has relied on the assumption that a person is a 'knowing' self, fully and consciously aware of him-/herself. In contrast, Sibley suggests that there is a part of the self that is unknown (and unknowable) – the unconscious part of the mind (see also Pile, 1996). Hence, we introduce the 'unknowing' subject, who doesn't (fully) know his or her own self. One thing that is expressed through this unknowable element of the unconscious, suggests Sibley, is the desire of human subjects to maintain a sense of order in their environment. The implication is that people have an (unconscious) desire to exclude those things which are 'unlike' themselves from their surroundings. This includes things like dirt and faeces, which are seen as 'unclean'. However, those things which are unlike, or 'other' to, the self may also come to include those *people* who are different, for example those having a different skin colour or being of a markedly different social class. Sibley suggests that we try to maintain 'purified' spaces free from defilement by dirt and 'other' people. This purification involves maintaining a physical and psychic distance on a variety of scales. At the scale of the body, for example, people feel that it is important to wash regularly to purify themselves of dirt and sweat (indeed, this may become a compulsive disorder for some people), while at the scale of the nation-state, exclusion may become politicized in the legislative creation of (for example) immigration laws regarded as protecting the 'purity' of those within the state (see section 5.3).

Returning to the scale of the domestic home, the desire to maintain purified space may be evident in two senses. Firstly, the opposition of inside (the home) and outside

Figure 5.1 Cleansing space

Source: Digital Imaging © copyright PhotoDisc, Inc.

('the big bad world') acts to reproduce a sense of a purified private home space as opposed to exterior public spaces. Controlling what and who enters the home is an important way of maintaining order, made evident when the home's privacy and order is defiled by burglars, through the often experienced feelings of a disgust which can only be assuaged by cleaning (Chapman, 1999). Secondly, we can recognize that there is spatial differentiation **within** the domestic home (as well as between the home and the outside). There may be expectations that muddy boots are not allowed in certain parts of the house, that food shouldn't be consumed in bedrooms, that the family pet isn't allowed upstairs or that children have to be in their bedrooms between certain hours (see also Chapter 9). Moreover, certain rooms may be considered as suitable to be seen by selected outsiders, whereas other rooms remain off-limits to all but the occupiers. These 'unwritten rules' of how the domestic home can be used are clearly the result of a complex social negotiation between different individuals. For example, adult ability to regulate the timing and spacing of children's activities (like setting a bedtime, or stipulating how much television can be watched) means that a child's experience of home is not always one of doing as they want, but one of behaving in such a way as an adult expects them to. This may apply even to the bodily postures and actions of the child – children are socialized by the parental expectation of them acquiring 'good' table manners, for instance, and not putting their feet on the chairs!

■ Exercise

Take some time to think about your own experiences of domestic home places, perhaps as a child or teenager, or as an adult, whether living on your own, or with other adults or children, or perhaps as a parent. How is (or was) the home differentiated internally? Think beyond the way that some rooms are clearly used as kitchens, bedrooms and the like, to thinking about the ways in which you and the people you live(d) with use(d) and experience(d) those rooms. Are or were there particular parts of the home that are associated with particular sides of some of the polar oppositions suggested above? What is or was your own experience of the home? How might that differ from the experiences of other people in that domestic arrangement? Can you say something about the sorts of power relations which are or were evident in that arrangement?

In the final part of this section, we want to explore the ways in which home places can become ambiguously meaningful in ways that imply that simple interpretations of home places become increasingly untenable. In her book *Yearning: Race, Gender and Cultural Politics*, bell hooks (1991) discusses her own childhood experiences of a racially segregated society in the USA. She writes about the way that black women constructed their homes as places of care, nurturing and retreat from a harshly racist society in which most of them also worked outside their own homes, in domestic service at the homes of white people. She argues that this assertion of home by black women, when viewed against a history of slavery, poverty, racial discrimination and domestic servitude, illustrates the way that home-making can have radical political potential. Home-making can potentially be a way of insisting on having a stake, having a (geographical and social) place. Making a home, even a fragile and temporary one, was for these women an act of resistance, which hooks claims was a way of regaining their 'subjectivity' (their personal human identities) in a society which tended to categorize them oppressively by gender and ethnicity as 'women' and 'black'. For hooks, identity is something which has strong links with home places. What is particularly interesting here is the way in which the conventional meanings attached to the domestic home – a place of withdrawal and privacy – are confirmed here but in a radically different way. For the black women of bell hooks' childhood, domestic work in their own homes became an empowering activity, in a way that the same work in their employers' homes was not (something Martin, 1984, picks up on in her analysis of household rituals).

However, a note of caution is necessary here, in that we need to be careful not to allow this association between women and domestic work to appear 'natural'. Instead, we need to see that domestic reproduction and home places can be linked to a complex range of meanings – places of rest and resistance at one and the same time. The powerful political potential of home places is certainly something which is recognized by governments. There are numerous cases of the destruction of informal dwellings and shanty towns on the margins of cities in 'developing' countries. Similarly, the temporary erections of the homeless populations on the streets of cities in the (so-called) 'developed' world have frequently been demolished, while 'squatting', even in unused buildings, is feared by the authorities on the grounds that it challenges the notion of private property ownership. The 'scandal' of the existence of unused housing stock in countries like the UK, when there are many individuals and families who are homeless, has been challenged by the formation in cities such as Nottingham, Brighton and Manchester of a sort of squatters' 'estate agency', which provides details of squattable buildings to potential home-makers (*Guardian*, 6 August 1999, 10). 'Home', in such cases, becomes a highly politicized concept, bound up with and confronting deeply held ideas about property and propriety (that is, who can own what and who should live where). From the perspective of the powerful (the wealthy, property owners, the government), squatting may be considered 'improper' because it involves people with no conventional ownership or tenancy rights making their homes in the property of others.

For some geographers, then, it has become essential to recognize that 'home' does not have a single, clear, fixed meaning for every individual, but that people's experiences of domestic home places are highly variable across age, gender, class and ethnic divides. We can start to see how the strident phrase 'An Englishman's home is his castle' – often used to assert the independence and privacy of a person in their house – is both a gendered expression and implies an ability to exclude (which may have repercussions at larger scales as we will see later). Ideas that men are inevitably the 'masters of the house' or the 'breadwinners' may potentially act as a screen for domestic inequality or abuse. Certainly, to be regarded as 'homeless' (as large numbers of people are in contemporary societies) has many negative connotations and very real insecurities. However, we need to balance this with the idea that having a home is not necessarily always entirely positive.

■ Reading

David Sibley's (1995) *Geographies of Exclusion* is a very readable account of some aspects of psychoanalysis in human geography. Chapter 6 ('Spaces of exclusion: home, locality, nation') is probably the most relevant for this chapter. Excerpts from bell hooks' (1991) *Yearning: Race, Gender and Cultural Politics* can be found in a collection of readings *Undoing Place? A Geographic Reader* edited by Linda McDowell (1997). The first section of this reader, 'Homeplace', consists of series of readings which can be used to destabilize comfortable notions of home places. Additionally, Chapter 2, ('Familiar places . . . home thoughts') of Pamela Shurmer-Smith and Kevin Hannam's (1994) *Worlds of Desire, Realms of Power* extends and adds to this process of destabilization.

■ 5.3 Exclusion, territoriality and national identity

In the previous section it was suggested that an important aspect of making and sustaining home places is the ability to exclude other people from that place. We want to extend this idea a little further here by considering the concept of **territoriality**. This is a concept whose origins lie in the environmental psychology discussed in Chapter 3, wherein researchers suggested that humans develop an understanding that certain spaces 'belong' to particular people. This may mean that people develop a desire to hold and defend a piece of space. Territoriality is something which becomes apparent at different scales, with, for example, the need for personal 'defensible' space around one's body and the bounded 'nation-state' both understandable as its manifestations. In Chapter 3, mention was made of 'ratomorphic' behaviour (the suggestion that some aspects of human behaviour and psychology is similar to those of other members of the animal kingdom), and territoriality is one of the psychological concepts which has been studied in experiments with rats, which show increasing amounts of aggression as they are placed in increasingly crowded conditions. Similarly, it has been suggested that humans become stressed when they see the boundary of 'their' space being invaded.

The notion that humans behave like other animals is highly questionable (see Gold, 1982), but the concept of territoriality does allow us to develop our critique of the conceptualization of home outlined in Chapter 4. Territoriality is undeniably an important aspect of home, implying a sense of control and belonging which relies on the ability to exclude others from that place. Territoriality implies the existence of unequal social relations at different geographical scales, in that some people claim the right to control who or what is allowed in their territories, while other people and things may be excluded from those territories. The existence of boundaries is also implied, to mark who and what is inside or outside a territory. What these points demonstrate is that territories (like states) are not things that are natural or fixed: rather they are created by people and are subject to variation in space and time. The concept of territoriality can be used to dispute the positive associations of belonging or attachment to place in at least two ways. Firstly, we can think about the creation and maintenance of different types of border, and secondly, about the way that some groups desire the exclusion of other groups from particular places. These ideas are elucidated by considering case-studies at a variety of scales, from the intra-urban to the international.

Notions of territoriality have often been evident in the maintenance of gang 'turfs' in the city. In the early 1970s there was a noticeable rise in incidences of gang warfare in various urban areas of the USA. Ley and Cybriwsky (1974) examined this issue in the Monroe area of Philadelphia, where the growing violence and graffiti linked with 'turf wars' between gangs of poor young males was seen as a serious social concern. Their research demonstrated that the gangs' use of graffiti was very much a part of their creation of territories, and their associated group identities. Graffiti, consisting of gang names and symbols as well as threats, can be understood as a way of confirming and communicating a gang's identity, giving gang members a sense of belonging to a gang which itself 'belonged' to a particular spatial area (generally a particular set of blocks within the city). Making graffiti, then, can be territorial behaviour. In this context, graffiti was intended to communicate something to two groups of people; figures of authority and rival gangs. Firstly, then, making graffiti was an act of resistance to the standards of behaviour and identities prescribed by, for example, the police and 'decent' society, particularly as these young males were what we might now describe as socially excluded (i.e. with little hope of work or money). New forms of identity arose from this resistance, based on being 'cool' or 'bad', which implied behaving in a way which would upset people in authority, while at the same time claiming a territory as their own place to be who they wanted. It was regarded as particularly cool to draw graffiti in challenging places – highlights being on a police car and on an elephant at the zoo. Secondly, graffiti was an expression of territoriality between rival gangs; part of the identity of gang members was to be very defensive of the gang's territory and aggressive towards members of other gangs. The result of this was a series of street fights, shootings and other violence. However, Ley's research showed that incidents of violence were more common between gangs whose territories were geographically close to each other (80 per cent of recorded violent incidents were between neighbouring gangs).

Figure 5.2, taken from Ley and Cybriwsky (1974), illustrates how the number of violent incidents declines with the distance between gangs. The key idea here is that the violence was part of a struggle over the boundaries between territories:

Figure 5.2 Turf proximity and inter-gang aggression, North Philadelphia, 1966–70

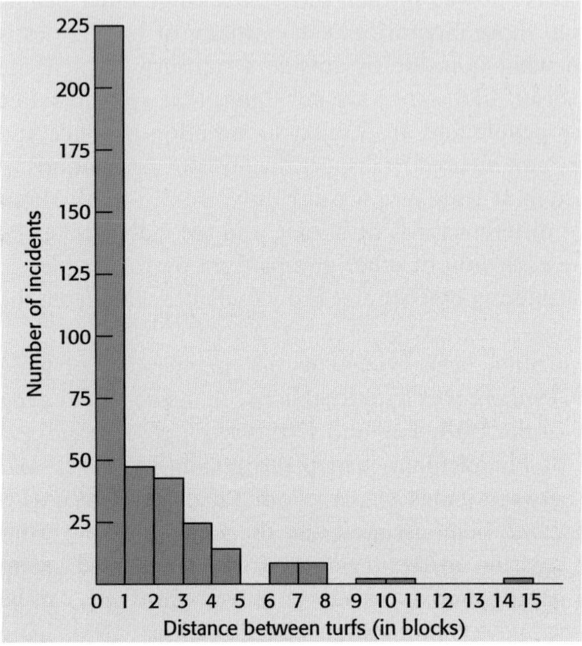

Source: Based on Ley and Cybriwsky (1974)

fights were about marking territories, defining boundaries, and maintaining gang identity; so, clearly, violence was more likely between gangs which shared a boundary. Linked to this suggestion is the distribution of graffiti across the gangs' territories. Graffiti tended to be concentrated at the boundaries of territories (see Figure 5.3). What we can say about this graffiti is that it was being used by gang members to communicate the identity and territoriality of their own gang to neighbouring gangs; it was a form of symbolic aggression and threat. In some cases, gang members would deface the graffiti of rival gangs, and this can be seen as part of the ongoing disputation of boundaries and territories. A further point worth making about this is that as a result of the conflict and violence between gangs, gang members became afraid to leave their own territories in case they were attacked in the territories of neighbouring gangs. This fear extended to those who were not even members of a gang. So territorial behaviour in some situations severely restricted the movements of people, and created places to be afraid of, or afraid in. We return to consider these geographies of fear later in this chapter.

Figure 5.3 Intensity of incidents between rival gangs

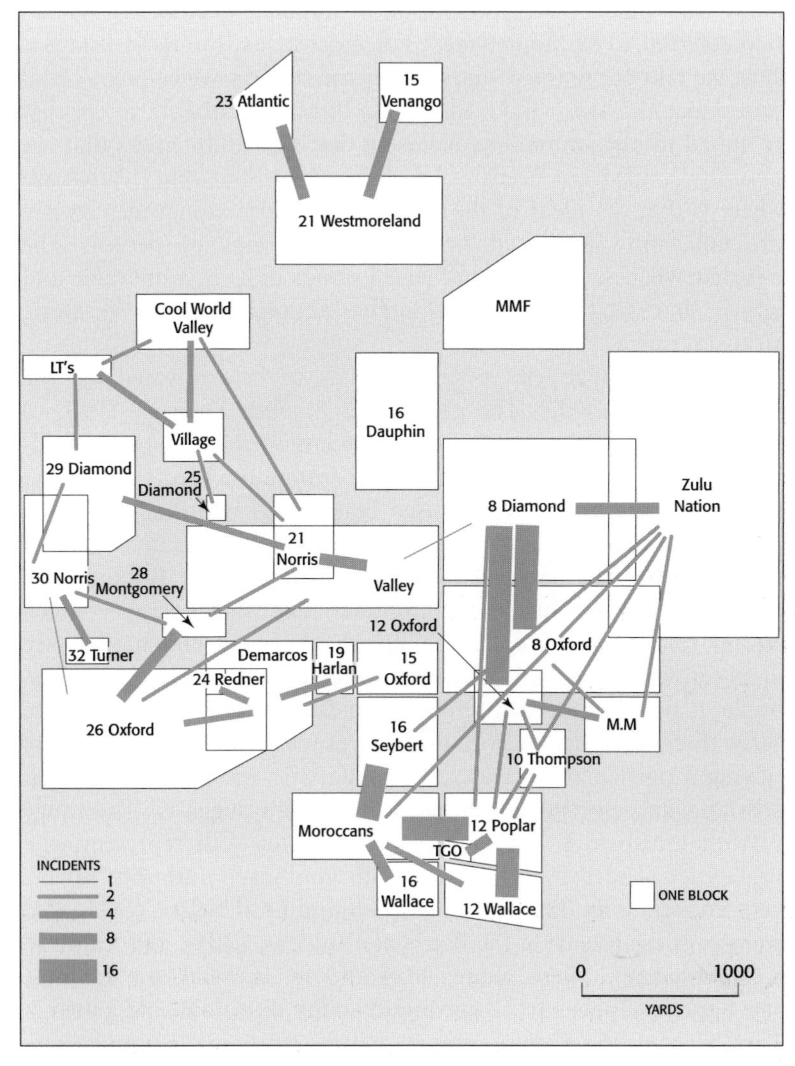

Source: Based on Ley and Cybriwsky (1974)

■ Reading

Ley and Cybriwsky's (1974) paper ('Urban graffiti as territorial markers') makes interesting reading here, though we might also encourage you to look at Cresswell (1992) for a more recent case study of how people use graffiti to create a different sense of place, and Davis (1991) for a lively account of the importance of gang territories in Los Angeles. Of course, knowing about these spaces and their boundaries is important for those charged with policing – something picked up in Herbert's (1997) excellent account of the territoriality of policing in Los Angeles.

Yet if territoriality is important on a microscale (e.g. in the home or on the streets), it is equally relevant in the consideration of **national identity** – after all, countries are often referred to as 'homelands'. For most of us, the nation-state is probably something we take for granted – it is simply the country we belong to (and others don't), the product of history and tradition. In turn, we probably accept that our identities are linked to our nationality, believing that we exhibit traits that are distinctly and naturally English (or French, American, Australian, etc.). What we want to suggest here is that the **ideas** of the nation-state and national identity are things which are constantly **produced** by particular groups of people. The implication here is that what the state is (and who belongs in it) is contestable and subject to change over time, with notions of distinctive 'national identities' breaking down under critical scrutiny.

A key idea here is that loyalty to a nation needs to be cultivated via the propagation of certain place myths. The geographer Stephen Daniels (1993, 5) argues that national identities are 'co-ordinated, often largely defined, by "legends and landscapes", by stories of golden ages, enduring traditions, heroic deeds and dramatic destinies located in ancient or promised home-lands with hallowed sites and scenery'. We can thus start to explore what can be thought of as mythical sets of ideas about nation-states and national identities. In the case of Britain, such myths are associated with things like the 'traditional' English landscape, which provides a backdrop to the evolution of the British character and nation (see Chapter 7), and a supposed national character (behaving in a 'gentleman-like' manner, for example, maintaining a 'stiff upper lip' and an attitude of paternalistic tolerance). Similarly, there is a range of icons and figures which is supposed to stand for Britishness, giving a particularly aristocratic, militaristic and heroic viewpoint on the story of Britain and Britishness, in the way Daniels suggests. The range includes royalty (with Elizabeth I, Queen Victoria or Henry VIII representing a supposedly natural order headed by a monarch with God-given authority to rule over their subjects), important military figures like Admiral Lord Nelson (celebrated for his naval victory over the French at the Battle of Trafalgar, 1805), and 'cultural' figures (such as Shakespeare). These figures may also be fictional, for example Britannia (a female figure symbolizing the unconquerability of Britain, and hailed in the triumphal 'Last Night of the Proms' concert – 'Rule, Britannia! Britannia rule the waves / Britons never, never, never will be slaves'), and John Bull, the figure of a stocky, patriotic and rather narrow-minded farmer dating from the early eighteenth Century, and symbolizing a British attitude of fierce independence from interference of any kind. (In Figure 5.4, from *Punch* magazine, 1847, John Bull is reprimanding the politicians to whom he has to pay taxes, for wasting their time!) Of course, Welsh, Irish and Scottish readers of this book (not to mention people from Cornwall, the Isle of Man, etc.) will be deeply aware that these are mainly English figures, and that the construction of Britishness has often involved the suppression of Celtic myths and images, for example (something we return to below).

Figure 5.4 John Bull and time-wasting politicians

WHIG PROGRESS.

John Bull. "WHAT! WASTING YOUR TIME, AS USUAL. PRAY, MASTER JOHN, WHAT HAVE YOU BEEN DOING ALL THIS SESSION?"
John (whimpering). "NOTHING, SIR."
John Bull. "AND WHAT HAVE YOU BEEN ABOUT, MASTER MORPETH?"
Morpeth. "HELPING JOHN, PLEASE SIR."

Note: The Whigs were a political party succeeded by the Liberals.
Source: *Punch*, 1847 © Punch Limited

At the same time, we can consider how the idea of national identity is also inscribed in particular places (and landscapes). Specifically related to some of the figures mentioned above are places like Trafalgar Square in London (Nelson) and Stratford-upon-Avon in Warwickshire (Shakespeare's birthplace), and there are many 'royal' buildings and landscapes dotted throughout the south (something documented in Taylor's (1991) mapping of places featuring in photographic collections of images of Britain, see Figure 5.5). More recently, the Millennium Dome in Greenwich was constructed as a place for the celebration of British values at the end of the second Christian millennium. These places have often become focal points for tourism, for British people perhaps attempting to obtain a sense of themselves and their historical provenance from these iconic sites/sights (Urry, 1990), and for foreign tourists who expect to be able to understand or absorb a stereotypical 'Englishness' or 'Britishness' by being in such places. In some ways, then, the idea of national identity can be thought of as an attempt by certain groups of people to promote their sense of place as the 'correct' or 'official' version to which everyone should subscribe. The idea engenders concepts such as a common history (or narrative of the emergence of a nation-state), common aims and ideals, a common character and shared moral values. What we need to do here is to consider some of the downsides of the assumption implicit within the idea of national identity, that a nation consists of a more or less homogeneous group of people who

Figure 5.5 Landscapes of Englishness

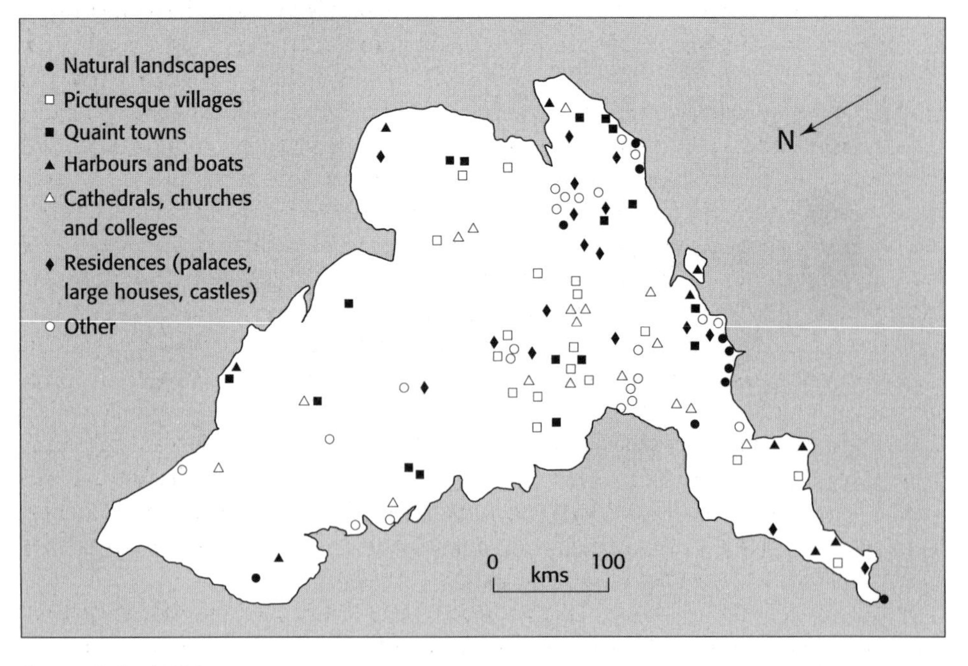

Source: Taylor (1991)

have a natural homeland within a defined and bounded geographical area. Following some of the ideas raised in previous sections of this chapter, we need, perhaps, to think in terms of inclusions and exclusions.

To begin with, we can use the case of national identity to criticize further the assumption of Chapter 4 that attachment to place, the creation of home, is wholly and simply a 'good thing'. The idea of a group of people having some sort of a homeland (or motherland, or fatherland) to which they belong through ancestral attachment may sound appealing, but we need to remember that as soon as we try to say that one group of people 'belongs' somewhere, there is the implication that other groups do not. Hence in the 1930s and 1940s, Jewish people (amongst others) in Germany were systematically persecuted in the genocide of the Nazi holocaust, in an attempt to try to racially 'purify' what was thought of as the German 'Fatherland'. In more recent times, the break-up of the former Yugoslavian state into a number of smaller states has seen violent conflict centring around which ethnic groups do and do not 'belong' in a particular piece of territory, while at the same time illustrating the arbitrary nature of the nation-state (i.e. it is a spatially and temporally specific human construct). This is not to deny its importance as a centre of human meaning. Indeed, creating a focus for collective displays of emotion and belonging is part of the purpose of the state. But at the same time, we need to be aware that attachment to home can develop into bigotry and hatred against those who supposedly do not belong.

Yet to a great extent, the idea of national identity also ignores the way in which the people living within a particular country are inevitably cross-cut by a great range of intersecting **differences**. By looking more closely at these people, we can immediately begin to see how they can be differentiated along the fracture lines of geography and ethnicity. Taking Britain as an example again, it has been suggested that what is posited as a British national identity is more typical of England than of Britain as a whole. Welsh and Scottish identities, for example, are often marginalized in the assumption that stereotypes of English history, culture, 'character' and landscape speak to (and for) the whole of Britain. But we cannot even say that there are distinctive English, Scottish or Welsh national identities, as within each of those areas there are significant regional differences playing an important role in how people living in them think about themselves (see, for example, the discussion of northerness and southerness in Chapter 7). This situation, as with all narratives of national identity, arises from a specific historical context. In the case of Britain, it is that powerful groups have conventionally been located in England, rather than Scotland and Wales (and, within England, in the South). National identities are thus forged by particular geographical relationships, meaning that some places become core and other places are marginal to the constitution of what is taken to be national identity.

As the atrocities perpetrated by the Nazis demonstrate, questions of nationhood often coincide with those of ethnicity. For better or (often) worse, nations are often constructed around stereotypes of particular 'racial' or ethnic types. This has a significant bearing on the life-chances of the different ethnic groups seeking to make a particular nation their 'home'. For example, Britishness has conventionally been associated with 'whiteness', contributing towards the social, economic and political marginalization of black and Asian people (despite the way Britain is represented as a multicultural society). The idea of national identity, then, can be associated with a 'mythical sameness which denies a place to many who are already there' (Sibley, 1995, 110). The issue here is that because of the stereotype of the British person as a white person, it is possible for racist groups in Britain to suggest that a person cannot really be British (that is, they cannot belong) unless they are white – even if they were born in Britain. They may consequently argue that there is a need to preserve the 'purity' of the British racial identity through the exclusion of 'other' ethnic groups. As with the purity of the nation state itself, this notional racial purity (as in fascist Germany of the 1930s and 1940s) has to be mythical, and is itself part of a wider set of processes involved with creating exclusive home places. All groups of people ultimately derive from mixtures, so that white British people emerge from a turbulent history of cultural hybridization involving the mixing-up of, for example, Scandinavian, Mediterranean and Celtic identities, despite what they might tell themselves about their unity and 'purity'.

To an extent then, the assumption or ideal of a national identity is associated with a denial of difference. Difference is subsumed under a myth of common history, culture and identity. But the truth is that 'nations are not really as solidly "placed", well-bounded or internally unified as we imagine them to be' (Hall, 1995, 183). The result is that the mythical story of a unified national identity is clearly open to challenge; we continually witness the assertion of many different, complex and constantly evolving identities related to place and ethnicity (as well as gender, sexuality, etc). In this case, why has the idea of national identity become so important? One answer is that national identity arises from the needs of powerful groups in society to maintain their control over weaker groups – national identity can be seen as a mechanism for the reproduction of unequal power relations, exploiting the human desire to belong. Looking at the language used by politicians and the press thus provides important clues to the construction of national identity. References to 'the British people' in political speeches, for instance, give the impression of a national unified voice, a sense of belonging to a national cultural community with common objectives and purposes. The politician claims to be able to speak 'for the people', and in so doing assigns a common identity which overrides the existence (indeed the possibility) of dissent and difference. Other voices are thus marginalized. In March 1998, for example, Peter Mandelson (then the Labour minister with responsibility for the Millennium Dome) stated – in relation to the Dome – that '60 per cent of people showed tremendous confidence that we will get it right. That is a *typically British attitude*, and their confidence will not be misplaced' (emphasis added). Similarly, Michael Portillo, speaking as a Conservative MP, once claimed that 'We need to assert the value and the quality of the *British way of life* and of British institutions . . . . We are proud of *our* history, proud of *our* language, proud of *our* culture, proud of *our* military skills' (emphases added). Note how the repetition of the word 'our' attempts to unify a variable population behind crude nationalistic sentiment, in the insistence that it can be unproblematically said that there is a typically British way of life. The construction and assumption of a shared set of values, morals, desires and purposes, and a shared pride in a common history, maintains the dominance of those who are powerful in society; if people feel that they are part of a common trajectory towards a common goal they are less likely to question dominant structures of authority (such as party-political government, or policing and legal systems).

When we consider the idea of nations, we are therefore entering into a complex set of debates about the manipulation of place identity in the interests of particular groups. This might make us think about the importance of national identity as part of a system whereby the political status quo is legitimated and reproduced. This can be seen in things like the British monarchy, which supposedly unites its subjects behind a common figurehead, and the Houses of Parliament, representing the birth of democracy. Having a monarch legitimates a whole system of graded inequality in society, which is presented as a 'natural order' because some people are simply born better than others. Using the Houses of Parliament as a symbol of British democracy, and in doing so suggesting that British democracy is the best system of government possible, acts to hide the limitations and faults of such democracy, and the social inequalities and injustices it actually reproduces. National identity, it can be argued, is a mechanism which pacifies a population by naturalizing (and hiding what is wrong with) the status quo while reducing the possibility of challenging it.

So a key driving force behind the existence of the idea of national identity is its political usefulness in situations of real or imagined threat to the nation-state. These threats can be categorized as external or internal. Externally, national identity has been mobilized in the defence of the 'British way of life' against, variously, immigration, war and the European Union. British people are also, at various times, invited by politicians to condemn specific groups, like single mothers, New Age Travellers, bogus asylum-seekers, etc., whose existence challenges dominant sets of values and lifestyles, and threatens what those in power would like people to regard as 'normal' (see Chapter 8). As David Sibley suggests, 'It is convenient to have an alien other hovering on the margins' (1995, 110), an 'other' who can be blamed for economic or social problems while removing attention from those in power. While suggesting that the idea of national identity is part of a marginalization of less powerful groups by more powerful ones, that is, of acting in a destructive way towards diversity, we should remember that the process of constructing the myth of a national identity is also a productive one. By this, it is meant that a particular version of a country's history is built up and used in particular ways. The history of the UK, for example, is often narrated as consisting of resistance to invasion (although the rhetoric of invasion has been employed by right-wing politicians in relation to immigration of black and Asian people into Britain), a series of great battles and a heroic ascendancy to world power through the creation of a great empire and the birth of the Industrial Revolution. This militaristic and imperialistic version of the history of Great Britain, while presenting a particular vision of a British character and destiny, hides a great range of alternative and 'quiet' histories – often including those of the poor, women and people of colonized parts of the world. Dominant histories, reproduced as they are in cultural forms such as the many Second World War films pitching the heroic and 'good' British against the 'bad' Germans, become national myths. Often these are factually inconsistent (for example, in their denial of the role played by Indian troops in the war, or in their unquestioning acceptance of British moral uprightness) and contribute to broader ideas about what other groups of people are like. For example, scripting of German identities around notions of aggressive territorial expansion appears to have influenced British attitudes towards its stewardship of the European Union.

History, then, is not a matter of truth and fact, but is a narrative forming a vital part of the construction of national identity, told from the perspective of those who are telling the story. We can talk about a plurality of histories in the same way as we can discuss a plurality of geographies, emerging from differently situated perspectives of the world. We might then summarize this section with a quotation from Stuart Hall (1995, 185) who argues that 'a culture is never a simple, unified entity, but always has to be thought of as composed of similarities and differences, continuities and new elements, marked by ruptures and always crosscut by *difference*'. National identity can be seen as a myth constantly produced in a range of ways and for a variety of reasons. This is not to say that a sense of national identity is unimportant; it clearly affects people's attitudes, behaviour and experiences of the (home) places they live in. But it may also involve the production of a sense of belonging which excludes many identities, histories and places in favour of others. In this way, the idea of national identity is hugely important, despite its mythical nature, because it has very real effects on the way people perceive and interact with their surroundings.

■ Exercise

Norman Tebbit, a right-wing UK Conservative Party politician, once suggested what became known as the 'Cricket Test'. The idea was that as national sentiment often crystallizes around support for national sports teams in international competition, a cricket match between England and Pakistan could be used to differentiate those Asian people in Britain who were 'really' British (because they would support England) from those who were not (they would support Pakistan). What can we learn from this episode about the nature of national identity? How does support for national sports teams relate to the ideas about national identity discussed above?

■ Reading

Stuart Hall's (1995) chapter 'New cultures for old' in *A Place in the World* (ed. Massey and Jess) expands on some of the ideas raised above, and also explores some other avenues of thought in relation to changing cultural identities. Chapter 6 of John Urry's (1990) *The Tourist Gaze* discusses some of the ways in which the idea of national and local heritage is constructed in places associated with tourism.

■ 5.4 Geographies of fear and anxiety

The preceding sections of this chapter have sought to show that ideas about home and attachment to place are far from straightforward, sometimes involving antagonistic relationships between contrasting senses of place. In this section we want to think more explicitly about how this produces fear and anxiety by focusing on the way in which some places are experienced as frightening or threatening. In Chapter 4, we mentioned Tuan's idea of 'topophilia' (love of place); here we consider the opposite – **topophobia** (fear of place). Tuan (1979), as a humanistic geographer, considered negative experiences of place in his book *Landscapes of Fear*. This eloquently described many of the experiences of fear and dread associated with different types of settings (e.g. the city, the 'wilderness', or simply 'the unknown'). Tuan argues that fear, as with emotions of belonging, is a fundamental human experience; fear can be associated with particular places (both specific places and types of place) in the same way that other places are associated with pleasant experiences. Yet fear or anxiety are emotions which are generally associated with being away from home, in places where you do not feel that you belong. This feeling of unease often results from a sense that the place belongs to other people in some way. Straight away, we can see that places are ambiguous in the way in which they can simultaneously be experienced by different people as places of belongingness and of frightening exclusion. Of course, the same idea applies to the idea of the domestic home, experienced simultaneously as a place of relaxation and work, comfort and frustration, power and subordination, by different members of the domestic household. At the same time, our supposedly safe and sacrosanct home-places may also be the focus of fear for other reasons – the very way in which the home is felt to be sacred creates fear of the potential violation of the home's boundaries (Chapman, 1999).

This fear may be exacerbated at night, a time when the streets are often imagined to be populated by threatening groups who might invade the home; the burglar who breaks down the door, the murderer who strikes under cover of darkness. Moreover, a range of evidence appears to suggest that this fear of the night takes different forms in the town and the country (see Schlör, 1998). This demonstrates the complex relationships that exist between urbanism, alienation and anxiety, something also highlighted by Tuan (1979). As he shows, cities are often understood in conflicting terms. One understanding or image we have of the city is that it is a place of order and human control over nature, something which is linked to an idea of a stable and harmonious society. The city, according to this idea, should be a place of safety. However, some experiences of the city directly contradict this ideal, and there is an alternative set of ideas about the city which represents the urban as unnatural, disordered and even sinful (see Chapter 6). There is a constant tension, then, between a desire (for most people at least) for the city to be a safe and ordered place, and the reality of our experiences of masses of people living in chaotic and confusing (if sometimes vibrant and exciting) urban areas.

The potential for fear and/or anxiety is thus ever-present and Tuan's work attempts to break this fear into a number of constituent parts, including (amongst others) fears of noise, fear of other people and urban complexity. The first of these concerns the constant presence of chaotic noise in cities which can be a significant contributory factor to feelings of stress and anxiety (at the same time as potentially contributing to an atmosphere of stimulation and excitement). Urban areas, particularly city centres, tend to operate against a wall of human and machine-generated noise – talking and shouting, the ringing of mobile phones, shop and street music, traffic noise, newspaper and street vendors, road drills, emergency sirens and car alarms. Although we may develop ways of dealing with this surfeit of stimulation (see Chapter 3), the tension produced by the constant noise can be exacerbated by the other fears which sounds may remind us of. Police sirens, for example, may heighten the fear of crime. Traffic itself can be a constant source of frustration, whether you are stuck in a traffic jam, or trying to dodge the vehicles as you cross the road.

A second source of fear in the city is other people. Cities represent agglomerations of large numbers of people, such that one can never know all of the people one comes across (see Figure 5.6). Fear of unknown people (and anxiety in unknown places) and what they might do to us is therefore a common experience – you don't know who is going to turn out to be the mugger. Initial judgements are often based on appearance, however, so that while, for many, smartly dressed businessmen or women are unlikely to cause concern, those people who deviate from standards of 'normality' in some way may seem especially threatening or simply 'out of place' (see Chapter 9). Thus, groups of young people, people from ethnic minorities, beggars, people rooting through litter bins, drunk people or those muttering to themselves may be foci of fear. These 'other' people, already marginalized in contemporary society in many ways, may be further excluded by what is often simply a fear of the unknown, which is often linked to media representations and popular understandings of deviancy or difference that suggest that particular groups of people are more dangerous than is actually the case. There have been panics, for example, about the dangers posed by mentally ill people on

Figure 5.6 Fear of crime and the night: *The Peril that Lurks in the Dark*, Gustav Doré (1863)

the streets, while unease about particular ethnic minority groups has existed in many places. There is a paradox here that while there are all these fears surrounding people who are 'different' from the 'norm', the city is still often represented very positively as a 'cultural melting-pot', where difference is accepted and even celebrated.

We need to remember that what we think we know about the people we encounter in different parts of cities is based on 'stories' of the city. Inevitably, some areas might be associated with competing stories which, on the one hand suggest that it is a risky place to visit, and, on the other, that it is a vibrant place to be. Having said that, these stories do have important effects on our behaviour. For example, we might avoid particular places because of the fear engendered by their negative stereotyping as hotspots of crime, with the media acting as a key source of our perceptions of crime (see, for example, Smith, 1984). Much literature has discussed the differential effects that fear has on different social groups, exploring the ways in which those fears constrain the way that people use urban space. Studies of women's fear of crime (e.g. Pain, 1991; Valentine, 1989, 1992) have suggested that women will avoid certain places more than men and are more fearful of crime than men, despite the fact that in the UK men are more likely to be victims of street crime than women (who in fact are often more vulnerable in their own homes). Similarly, elderly people, people of certain classes or from ethnic minority backgrounds may feel constrained in their use of space by a fear of encountering dangerous people in certain parts of the city (and at certain times). It may thus be *perception* of risk which is important in considering how fear influences the use and experience of urban space.

Thirdly and finally, Tuan (1979) points to the influence of urban complexity in exacerbating fear and anxiety. The very size and complexity of cities may add to feelings of unease, as although we may know parts of a city very well, we are unlikely to know all areas well enough to feel 'at home' there. Consequently, the experience of being in the city can be very alienating (listen to late 1970s' songs like 'Strange Town' and 'Down in the Tube Station at Midnight' by The Jam, or 'Ghost Town' by The Specials). There is often a contrast in contemporary Western cities between those areas that are formally planned and structured, and those areas that seem to have grown haphazardly resulting in a bewildering pattern of streets and buildings. There is thus the potential for a fear resulting simply from getting lost in unknown places. Evidence of this type of fear exists in those English novels of the Victorian period, tending to be written by male members of the bourgeoisie, which take the reader into the 'underside' of nineteenth-century urban life – into the unplanned squalor of the slums, for example. Tuan (1979, 162) quotes Charles Dickens' description of part of London as an example of a fear of unknown, squalid people and places:

> a black, dilapidated street, avoided by all decent people; where the crazy houses were seized upon, when their decay was far advanced, by some bold vagrants, who, after establishing their own possession, took to letting them out in lodgings. Now, these tumbling tenements contain, by night, a swarm of misery. As, on the ruined human wretch, vermin parasites appear, so these ruined shelters have bred a crowd of foul existence that crawls in and out of gaps in walls and boards.

Note here the explicit contrast drawn between normal 'decent' people and the slum-dwellers and vagrants, described as various forms of repellent insect. The very vagrancy of the vagrants is sufficient to 'other' them in comparison with 'decent' sedentary life (see Chapter 9). It is possible to identify a form of environmental determinism at work here associating the people with the appalling accommodation they have to live in – to live in decaying places like this, it seems, is to decay morally into some form of sub-human life (Driver, 1988).

As you read through these ideas about the production of fear, you may find yourself thinking about your own fears as you move through urban space. As we will see in Chapter 8, attempts to make city centres safer (via policing and CCTV surveillance, for example) may target groups who do not conform to this ideal, aiming to circumscribe their ability to use public places. Perhaps we should turn this thinking around, and consider how threatening city life might be for the homeless, mentally ill or disabled who may have very real fears of intimidation and violence. Equally, in some cities, newly arrived immigrant groups may be perceived as threatening and dangerous – but for them the experience of being in new and unknown cities (or countries) is itself deeply scary and unsettling. Like us, you might be wary of the fact that many attempts to make cities safer seem to be based on the elimination of groups seen to be threatening from the perspective of white, middle-class, adult males (see Oc and Tiesdell, 1997).

A final point to make here is that fear and anxiety about the different types of people we encounter in the city has broader geographical consequences, being implicated in the development of residential segregation. For example, the middle classes have tended to separate themselves off from poorer or working-class people, and to maintain (physical and psychic) separation through the mechanisms of the housing market. Such class segregation has often been overlain by racial segregation (as immigrants from ethnic minority backgrounds have tended also to have lower-paid and lower-status jobs, at least initially). Of course, this segregation has both economic and social underpinnings, with immigrant groups having often clustered together for feelings of security and belongingness among people and cultural forms that they feel comfortable with. Indeed, there are many cities with areas which are noted for having a character which is related to a concentration of particular groups of immigrants and their descendants (such as the 'Chinatowns' which are often promoted by Western cities as cultural attractions, and black 'ghettos' – which aren't). The geography of race and racism is very complex, then, but in an era of globalization it is clear that a variety of different **ethnoscapes** are being created (see Chapter 2), each associated with particular feelings of desire and/or fear.

■ Exercise

As you move around the cities where you live or study, or that you visit, try to monitor your feelings of belongingness or fear/anxiety. What exactly is it about different parts of the city that results in these feelings? What is it about *you* that might influence these feelings? Are there particular groups of people who might give rise to anxiety (it is likely that for most people there will be)? Why? Finally, to what extent does this fear constrain your movements? Are there areas you would avoid, or only go to with friends? Again, why? Try to think through the same sorts of questions from the perspective of other people – for example, if you are male, how would a woman's experience be different from your own, or if you have a home, how would a homeless person's experience differ from yours?

Reading

We have drawn extensively here on Tuan's (1979) *Landscapes of Fear*, an extremely interesting and accessible read, focusing on fear in different categories of place. Two papers by Valentine (1989, 1992) are recommended for their focus on women's experience of fear (concerning 'The geography of women's fear' and 'Images of danger: women's sources of information about the spatial distribution of male violence', respectively). Rachel Pain's (1997) paper on 'Social geographies of women's fear of crime' develops similar themes. Finally, a classic reference on the social life of cities and the fear of strangers is Sennett (1977).

5.5 Rethinking humanistic geographies

This chapter has been largely concerned to cast a critical eye over some of the ideas associated with the humanistic geographies (and geographers) discussed in Chapter 4. In this, the final section of the chapter, we want to summarize some of the key ways in which these critical ideas can be used to develop two of the concepts at the heart of humanistic geography – namely, place and landscape.

Beginning with place, contrary to the humanistic suggestion that it is possible to uncover a true, authentic *genius loci*, the above discussion illustrates that places always have multiple identities. Different social groups engage with places in very different ways, so that places can be experienced in different ways according to a person's gender, social class, ethnicity and so on. Richard Peet (1998) and Stephen Daniels (1985) both argue (from very different standpoints) that there are several key problems with the humanistic conceptualization of place, particularly in relation to the idea that it is possible to uncover the 'true' meaning or character of a place. Peet suggests that this assumption of a hidden truth is nothing but a misplaced nostalgia, part of a process whereby the past is romanticized. The implication here is that the idea of unique and authentic spirit of place is selective sentimentality, 'a delightful yet dangerous way of cleansing the past' (Peet, 1998, 64). Why is this so seductively dangerous? We can suggest three (interlinked) reasons. Firstly, there is a denial of difference, meaning that it skims over the nuanced (and frequently individual) experiences of the people whose identity is partly constituted through being in a place. Instead, there is an assumption of a 'universal subjectivity' implying that all people share the *genius loci* immanent within a place. Secondly, the denial of difference and insistence on authenticity is linked to an important erasure (i.e. something is hidden or forgotten). This is the way that very different experiences of place are ignored in favour of the 'stories' and experiences associated with the more powerful social groups (who have more ability to define place meanings). This is seen, for example, in the way that women have been told that their (social and geographical) place is in the home. As Daniels argues, this means that the concept of 'home' carries very conservative overtones, in that it is reactionary of backward-looking in its desire to reproduce a particular social order.

Much humanistic geography, therefore, lacks a critical engagement with the power-laden differences of, for example, social class, gender, sexuality, ethnicity, and the resultant inequalities impacting on experiences of place. In some accounts, the humanistic valorization of home may condone poverty and injustice by failing to analyse critically different experiences of home (Daniels, 1985). This criticism is associated with a third criticism of humanistic approaches. This revolves around the way in which a phenomenological approach (see Chapter 4) seeks to excavate an 'authentic' experience of place which exists outside the realms of language. This is a difficult idea to grasp, but, in essence, some humanistic geographers would argue that language acts as a barrier that obscures what people 'really' feel about a place. A critical response to this has been to suggest that place meanings can never be understood independently of language, and in fact that meanings arise *through* language. For example, for Seamon (1979) to be able to discuss 'place ballet' requires us to be able to make something of the concepts of 'place' and 'ballet' before and outside of a humanistic engagement with place. As Daniels (1985, 147) puts it, it is a 'fallacy that a convention like language is an impediment to understanding'; instead, it is language that creates the concepts and meanings which are necessary to understand anything of the world. It is impossible, therefore, to approach the world with minds free of preformed concepts and theories, and for some the humanistic suggestion that we should try to is itself a theoretical injunction.

All this is not necessarily to deny the importance of place as a key concept in human experience, but it does inject some notes of caution lest we try to apply these concepts indiscriminately. Ultimately, the question of **evaluation** needs to be considered when we think about the usefulness of humanistic concepts such as place. To explain this, Peet (1998) describes how much of humanistic geography rests on a series of oppositions, such as inside vs. outside, belonging vs. not belonging, depth vs. superficiality, experiential understanding vs. scientific knowledge. In all of these oppositions the first term is valued over the second, and Peet contends that this valuation lays bare the very human theoretical construction which lies at the heart of the humanistic attempt to overcome language and theory. In short, this is a question about whether geographers' attempts to understand and know the world are always locked into a particular way of thinking that cannot do justice to the real complexity of the world. This is not a straightforward question to answer – and it is one to which we return in Chapter 10 – but by pinpointing the links here between language and theory, we might hopefully have made you think about the way the words like 'place' can be used emotively to support particular geographical ideas.

■ Reading

Doreen Massey (1995a) discusses the ways in which places have multiple and changing identities, as opposed to fixed immutable ones. Daniels' (1985) chapter 'Arguments for a humanistic geography' is a good summary of some philosophical criticisms of humanistic approaches (another interesting version is given in Pile, 1993).

Finally, we return to the concept of landscape. Following the lines of critique set out above, it can be suggested that a humanistic approach to landscape (as proposed by Edward Relph, among others) needs revision to consider the fracturing effects of power and difference. In Chapter 4, Relph's (1987) approach to landscape was noted as being a deliberate focus on superficial (visual) appearance, as this was taken to be the very essence of landscape. However, Barnes and Gregory (1997) challenge this understanding of landscape, arguing that it is not enough simply to study the visible form of a landscape. From their vantage point, attention needs to be paid to the ways in which human societies have created landscape (e.g. through exploitation for industry or agriculture, for example), necessitating an examination of the unequal relations which bring landscapes into being. This might involve looking at patterns of landownership, the aesthetic assumptions underlying the creation of formal gardens, or the **iconography** (shared symbolism) of particular architectural forms. Landscapes, then, are not just superficial visual expressions of relationships between society and nature, but the creative products of specific social contexts and power relations. They may thus be the result of historical and ongoing struggle, and indeed certain landscapes may come to symbolize struggle between different groups, or a struggle to construct national identity. In the UK, for example, moorland landscapes have been implicated in struggles between walkers wanting the 'right to roam' and landowners wanting to exclude people from their private properties. The landscape in cases like this does more than simply provide a backdrop: it can be thought of as playing a physical and symbolic role in the struggle (see Chapter 7).

Gillian Rose (1993) takes these arguments a step further in suggesting that the way we look at landscapes is structured by the way society is ordered. Her development of a feminist perspective on landscape encompasses the way in which looking at landscape is implicated with ideas about knowledge and control. She suggests that there has been a conflation of 'seeing' with 'knowing' in Western societies (so that for example, to 'gain an insight' is to know something; to be 'shortsighted' can imply a lack of knowledge). Hence, to gaze upon a landscape is to enter into a power relation with it where the gazer is attempting to gain knowledge, and hence possible control, over it. Moreover, Rose suggests that we can understand this powerful way of looking as a masculinist ordering of society and environment. Landscapes have often been represented as feminine in character (as, for example, 'mother nature'), and the way in which they are looked at **objectifies** them in the same way as women have often been objectified (as possessions, or as objects of desire). This, in effect, subjugates them, rendering them passive under the commanding eye of men. The dominant masculine gaze thus constructs the landscape as something which is known, controllable and both pleasurable and useful (see also Nash, 1996). The lesson here is that we need to be careful when viewing the landscape through the 'lens' of humanistic theory, and need to think about the possibility of alternative viewing positions. In the following chapters we go on to explore some further ideas about how social and cultural differences are played out in particular places (and landscapes), extending our remit to think about the **representation** of place. As we do so, ideas that place meanings are not just individual but social creations will become more apparent, presenting us with some different ways of thinking about these links between people and place.

■ Reading

The section by Barnes and Gregory on 'Place and landscape' in their anthology *Reading Human Geography* (1997) can be usefully contrasted with the references to Edward Relph in the previous chapter. Rose's (1993) arguments appear in Chapter 5 of her book *Feminism and Geography*, and excerpts can be found in Barnes and Gregory (1997) and Agnew, Livingstone and Rogers (1996).

■ Summary

This chapter has outlined some of the ways in which concepts derived from humanistic geography have been adapted by geographers to introduce new dimensions into the study of the relationships between people and places. A particular focus has been given here to the idea that places (and landscapes) are potentially experienced differently by different people. Major points to emerge here have been:

- That people are separated along many lines of difference (of gender, ethnicity, age, etc), influencing their feelings for (and fears of) certain places.
- That we can think about places and landscapes in terms of their embeddedness in networks of social relations (e.g. the unequal relationship that exists between men and women).
- That different people will engage with places and landscapes in very different ways, suggesting that it is difficult to support the idea of an immutable *genius loci*.
- That place meanings are constructed and mediated through language, rather than being inherent in any place.

six Imagining places

This chapter covers:

6.1 Introduction

While Chapters 3, 4 and 5 concentrated primarily on the encounter between individuals and their surroundings (via behavioural and humanistic ideas about place), this chapter will focus mainly on **collective** understandings or imaginings of place. In doing so, the notion that we understand places through socially constructed **mythologies** (or myths) will be foregrounded. In this context, myths will be defined as those powerful 'stories' which shape our physical (bodily) and imaginative journeys into both familiar and unfamiliar places and spaces. While it is important to note that these spatial mythologies are not universally shared (that is, different people have different sets of beliefs and understandings about places and spaces), we want to show that the existence of certain mythologies in particular social contexts is fundamental in shaping people's attitudes to particular types of place. A central idea that we will explore is that these myths are **relational**, setting the characteristics of one place (or region) against another so that one is imagined as a centre, the other as a periphery. Consequently some places are associated with wealth, style and sophistication, while others become imagined as sites of despondency, poverty and backwardness. In turn, the people who are imagined to live and work in such different environments also become imagined in terms of these opposed notions of centrality/peripherality (which translate into ideas about their social status).

As we will see, although the nature of these myths is often complex, these centre/periphery distinctions tend to endure over time. In many ways, this is connected to the unequal social relationships which allow these myths to become widespread, as those who occupy centres often control the resources which allow them to maintain differentials of wealth and status between themselves and those in peripheral regions. As we saw in Chapter 5, there are various ways in which those with the means to do so can exclude groups from 'their' space, casting them to the periphery. The powerful may thus employ a variety of means to sustain the (real and imagined) distance between themselves and the 'others' who are effectively treated as outsiders. In this way we will begin to see that although these myths are just that – sets of ideas – they do have 'real' consequences on the way people occupy (or are allowed to occupy) particular places. To emphasize this point, this chapter will explore the nature of a number of powerful mythologies which exist at different scales, highlighting what they can tell us about the people who produce the mythologies and the wider relationship between people and place. Before we do so, however, it is useful to clarify what we mean when we talk of 'place myths'.

6.2 Mythologies and geographical imaginations

In discussing mythology in this chapter, it is important to stress that we are not restricting our definition of the word to commonly held notions of (often heroic) stories, such as the Arthurian legends concerning parts of the UK, or the Ancient Greek tales of the Olympian gods. This is not to say that these myths are not worthy of geographical investigation; after all, such myths often play a part in the construction of national identity (such as the story of St George and the dragon – a tale that seemingly underpins the ideas of Englishness we discussed in Chapter 5). Besides, these stories of a fabled and fabulous heritage have often been used as the basis for tourism, and numerous museums, heritage sites and amusement parks are themed around such myths (Urry, 1990). Here, however, a mythology is defined as an often vague and diffuse way of **imagining** particular 'real' places and the people in them. Such myths may be considered as a set of 'stories' about a place, stories whose origins and characteristics are difficult to pin down but become widely known, and often accepted, as having some basis in truth. Inevitably, these stories or myths serve to **stereotype** a particular place or set of places by highlighting some of its characteristics in favour of others; in many ways, we need these myths to help us make sense of the world and our place in it.

It should be made clear that while a myth can be considered as a story, it should not be dismissed as either incorrect or unimportant. In the introduction to his book *Imagined Country: Society, Culture and Environment*, John Short (1991) quotes from Richard Hofstadter's (1955) *The Age of Reform*, where a myth is defined thus: 'By "myth" I do not mean an idea that is simply false, but rather one that so effectively embodies men's [*sic*] values that it profoundly influences their way of perceiving reality and hence their behaviour.' Accordingly, Short reasons that myths communicate across time, and, in effect, 'destroy' time. As he suggests, myths 'are messages passed through the ages and over the generations, kept fresh by use and reuse . . . they are of fundamental importance in how all societies "see" their physical environments' (Short, 1991, 3). In developing this argument we might further draw on Short's idea that myths tend to deal with abstractions, so that we can talk about generic myths associated with 'the city' or 'wilderness' or 'the Orient' rather than stories referring to specific places. Of course generic myths may be made up of stories about specific places and people, but these are generally used to support more general stories. This in itself emphasizes that it is unnecessary to consider the truth or falsity of myths (i.e. whether they bear resemblance to some 'reality'). Geographers, then, have tended to ignore questions as to whether these stories are fact or fiction in favour of critical analyses of their 'material' effects. This entails a consideration of how such myths become one of the very real ways in which people come to terms with (and change) their geographical environments, reducing unfamiliar people and places to familiar stereotypes of 'self' and 'other' (see Chapter 4).

As well as influencing the way in which people 'see' their surroundings, myths have two other important functions (as will be demonstrated by looking in detail at some case-studies). The first of these is that myths affect the ways that people behave in and engage with their everyday (physical and social) environments. The second is that the myths associated with a group of people can be examined in order to say something about the people themselves. Geographers might thus be interested in the ways in which a group of people living in a specific city (London, New York, Paris, etc.) imagine and mythologize the (abstract) 'city' in relation to their everyday urban experiences, as well as the ways they create and sustain mythologies about surrounding 'countryside' or 'wilderness' areas. For the city-dweller, the country may be something that is imagined as 'other' to the areas and places with which they are most familiar. Their experiences of the city and countryside which they encounter will occur in relation to these abstract myths of 'city', 'country', etc. One particular characteristic of such mythologies is that they tend simultaneously (and ambiguously) to consist of both positive and negative elements. Again, this is of importance in demonstrating that mythologies refer to imagined abstractions, but it also leads us to consider the complexity of the human imagination. That is, we can use the existence of mythologies to say as much, if not more, about the people who 'believe' them as about their abstract spatial referents (city, wilderness, etc.).

The bulk of this chapter, therefore, is devoted to exploring some mythologies associated with cities, wildernesses and the 'East' ('Orient'). Each concerns a widespread set of ideas about the characteristics of a category of places which are inevitably contrasted with other types of place through binary oppositions of self/other, centre/periphery, good/bad, order/disorder and so on. For a variety of reasons, the mythologies covered below are largely associated with Western, European societies, and they are frequently gendered (particularly masculine) imaginings. This should not distract from the fact that many different ways of imagining and mythologizing the world exist at a variety of scales, and that there are many other place myths that could be highlighted (something that we attend to in our suggestions for further reading). Nevertheless, the mythologies we focus on here have arguably been particularly important in shaping the everyday geographies of people in Western societies, and we hope that you will begin to recognize their importance and ubiquity.

6.3 Urban myths

On one level, urban myths are stories about the city, which, apocryphal or not, have become common currency. For example, in the early 1990s there was a widely known story about a 'yuppie' on a train talking loudly into a mobile phone, which, when requested by a fellow commuter to summon a doctor for a passenger suffering a heart attack, turned out to be fake. While amusing, this story implies something of the contemporary 'urban condition' – in this case both the (assumed) status bestowed by the use of a mobile phone, and the figure of the image-conscious yuppie preoccupied with conspicuous consumption. Yet this story is also relatively abstract: it could have occurred in any of a huge number of similar places or situations (perhaps you have heard of similar stories which you have embroidered with particular place-specific details before recounting to your friends or colleagues). This arbitrariness is not as important as the fact that we find this story worthy of telling, presumably because we find it entertaining and it helps us to make sense of the encounters we experience in our everyday lives. While the example of the mobile phone is, perhaps, limited to a particular time–space context, some writers (e.g. Williams, 1973; Shields, 1996) have been able to suggest the existence of stories of the (abstract) city which span larger intervals of time and space and have profoundly influenced our experiences of urban life.

Exercise

Before reading on, make a short list of what comes to mind when you think of 'the city' or 'the town' – not the specific cities or towns you are familiar with, but the abstraction; what is the nature of 'city' or 'town'? By doing this you should be able to start examining the myths of the city which influence your experiences of urban places you know.

A number of images have, no doubt, sprung to mind. Broadly speaking, and drawing on Short (1991), pro-urban and anti-urban myths can be identified. Although these are outlined separately below, it is essential to understand that the pro- and the anti- are inseparable; each takes form in relation to, or in opposition to, the other. Like the two sides of a piece of paper (can you imagine a piece of paper with just one side?), one is reliant on the existence of the other. This stresses that we are always compelled to try to make sense of one place (or object, person, etc.) by reference to what it is not – its 'other'. The implication here is that we can only understand things and places in terms of the ways in which they are *different* from other things and places.

If the city is imagined through antithetical notions of desire and disgust (i.e. through both positive and negative myths), it is perhaps useful to examine the basis of these myths. These are not solely recent inventions, having developed over hundreds of years. For instance, in the classical Greek and Roman periods, urbanism was associated with the idea of 'civilization', planting the seed of the myth that the city is the seat of culture, learning, government and civil order (Sennett, 1994). Since that time, the city has been an icon of progress, enlightenment and opportunity, a sentiment celebrated in the story of Dick Whittington, for example. According to this tale, the eponymous hero set out for London where, he had heard, the streets were 'paved with gold'. He ended up becoming Lord Mayor of London, of course, justifying this view of the city as being full of opportunities for those bright or lucky enough to be able to take them. Similar myths abound in other nations, of course, not least in the USA, where Hollywood's 'Dream Factory' reputation has inspired many to seek out the bright lights of the big city in search of fame and fortune. In the 1991 film *Pretty Woman*, the character of the street prostitute played by Julia Roberts apparently shows this is a routine occurrence when she is swept up off the streets by a wealthy businessman (Richard Gere). While there are many subtexts in this film (e.g. heartless businessman finds redemption in true love), the cliché of Los Angeles being a city where you can make your dream come true is one that underlines the pernicious myth of the city as a place of opportunity.

In part, this romantic view of city life is supported by the idea that cities are connected to what is happening in the rest of the world, opening a world of opportunities. From their earliest origins cities have frequently been centres of international trade and migration, global media and international politics, a strategic role that is sometimes imagined to be enhanced in an era of globalization (see Chapter 2). Further, the city is viewed as having dominion over the surrounding countryside. Thus, pro-urban myths exist not just in relation to anti-urban myths (see below), but also in (spatial) opposition to the imagined countryside. If cities are cultured and vibrant the stereotype of the ignorant and brutish yokel or 'country bumpkin' constitutes one side of the countryside myth. This stereotype contrasts sharply with that of the liberal-minded and educated townsman or woman, perpetuating the idea that rural folk are isolationist and technophobic in contrast with sharp-suited and quick-witted urban dwellers who keep their 'finger on the pulse'. To underline this, Short (1991, 43) paraphrases Marx and Engels in saying 'towns saved people from the idiocy of rural life'.

At this stage we must also acknowledge that there is also a pro-rural myth – that of the 'rural idyll' (see Chapter 7) existing in opposition to yet another set of anti-rural myths (see, for example, Bell, 1997, on ideas of 'rural horror'). Nonetheless, the connection between urbanism and order, progress, power and learning is widespread, sustaining other pro-urban mythologies. For instance, we might think about the myth of the city as a cultural 'melting-pot'. This stresses the very size of the city as providing opportunities for variety, social mixing and vibrant encounters between very different social groups. Because of this, the city may also be regarded as having a radical potential, where it is possible to challenge entrenched order and struggle for liberty and egalitarianism. The French and Russian revolutions, for example, had largely urban roots. It has been difficult, therefore, to specify a single pro-urban mythology if we consider that the city is associated with both order and revolution, and that both of those things can be thought of as positive. Neither are these things solely urban. The countryside (in Western Europe) has traditionally been imagined as an ordered, idyllic space, although the contemporary countryside is increasingly becoming a site of protest against entrenched social order. In Britain, for example, hunting with dogs and experimentation with genetically modified crops have recently been challenged by people from both urban and rural backgrounds, while demands for access to privately owned moorland and mountain areas have been made for many decades.

All this goes some way to demonstrating that myths are just that – myths – and exist as fragmented, ambiguous and often contradictory sets of ideas. For all that, they are still highly influential in shaping imaginings of different social and physical environments, becoming key reference points in the seemingly common-sense ways in which a society collectively thinks about something (the city, for instance). And, even taking into account the way these myths change over time, it is clear that many are highly resilient. The following exchange, in the early part of Jane Austen's *Pride and Prejudice* (1813), illustrates how remarkably similar the imagined city and country were in the early nineteenth century to the present. The (broadly rural upper-middle class) characters here are Elizabeth Bennet (the heroine of the story), Mr Darcy (the eventual hero), Mrs Bennet (Elizabeth's rather foolish mother) and Mr Bingley (friend of Mr Darcy and neighbour of the Bennets). They are discussing Elizabeth's fondness for studying people (subjects):

> 'The country,' said Darcy, 'can in general supply but few subjects for such a study. In a country neighbourhood you move in a very confined and unvarying society.'
>
> [Elizabeth replies] 'But people themselves alter so much that there is something new to be observed in them for ever.'
>
> 'Yes, indeed,' cried Mrs Bennet, offended by his manner of mentioning a country neighbourhood. 'I assure you there is quite as much of *that* going on in the country as in town. . . . I cannot see that London has any great advantage over the country, for my part, except the shops and public places. The country is a vast deal pleasanter – is it not, Mr Bingley?'
>
> 'When I am in the country,' he replied, 'I never wish to leave it; and when I am in town, it is pretty much the same. They each have their advantages and I can be equally happy in either.'

> 'Ay, that is because you have the right disposition. But that gentleman' (looking at Darcy) 'seemed to think the country was nothing at all.'
>
> 'Indeed, mamma, you are mistaken,' said Elizabeth, blushing for her mother. 'You quite mistook Mr Darcy. He only meant that there was not such a variety of people to be met with in the country as in town, which you must acknowledge to be true.'
>
> 'Certainly my dear – nobody said there were; but as to not meeting with many people in this neighbourhood, I believe there are few neighbourhoods larger. I know we dine with four-and-twenty families.'

Mrs Bennet's spirited yet comic defence of the countryside from charges of narrowness contributes here, ironically, to the widening of the gulf between (abstract) town and (abstract) country, and exemplifies the significance of such mythological categories in many social contexts. It is not that any of the comments made by Austen's characters are right or wrong, but that such a conversation, using the categories of 'town' and 'country', is held at all, which is important in the context of this chapter. The frame of reference has changed little over the last two centuries, such that the supposed contest between the supposed polarities of town and country makes regular appearance in public discourse. The exercise below illustrates this with excerpts from a newspaper cutting from the *Guardian* which, by highlighting the downsides of living in the village of Boothby Graffoe (a real village in the English county of Lincolnshire), draws on a basically pro-urban mythology while playing on rural stereotypes.

■ Exercise

Read the newspaper excerpt below, then take some time to think about the following questions. From what perspective is the article written (i.e. who is 'us' and 'them' in this context)? What specific aspects of rural mythology and urban mythology emerge from this excerpt? How could the article be challenged by drawing on alternative imaginings of the rural and urban? How do the mythologies drawn on in the excerpt relate to the 'truths' of urban and rural living?

I always think of Boothby Graffoe when the town versus country debate goes another round . . . . Sure, it's a pretty enough place, quaintish homes and lichen-covered farm buildings; hardly any traffic, it's safe for young children and the elderly; there are well-signed field-paths from which you can view spectacular sunsets over the Trent Valley.

But the peace that falls like a pall on Boothby Graffoe can bring [on] gloom. It's not merely funereal, it's sepulchral. There's no village store, no butcher, baker nor trendy scented candle-maker; no pub or off-licence. No Chinese or Indian caterer has sought to change eating habits, unchanged since Doomsday, by opening a take-away . . . .

> Somehow, a street . . . in a community surviving thanks partly to tele-cottaging professionals and weekending townies, surrounded by a wilderness of farmland much of which we no longer need to feed us, is seen as more honest, more virtuous than a city.
>
> Let's admit at once that you can compile a whole alphabet of horrors that accurately describes the depths to which conurbations have sunk. For a start, they're Alco-ghettos, Bestial, Corrupt, Druggy, Environmentally Dangerous, Fractious, Ghastly, Hazardous, In-Your-Face . . . .
>
> And just as many myths scarf the truth about city life as living in the country, away from the stress. But in spite of this, if we have to have this debate about where to live . . . surely cities have the ultimate merit of calling us out of ourselves in a way that villages rarely do.
>
> Cities, with their economic, social and racial mixes, stretch our tolerance and generosity; remind us in the swarm of culturally different faces that we belong to a bigger community than Boothby Graffoe. Or even Bayswater [an area of central London]. If we're lucky, cities smooth the corners of our personalities. In villages our lives are right-angled. And festering feuds thicken like weeds, diminishing lives.
>
> (John Cunningham, 'The trouble with Boothby Graffoe', *Guardian*, 12 August 1997, p. 7)

Confusingly, perhaps, both town and country are associated with particular ideas of desire and disgust, power and loss, order and disorder. Anti-urban myths, for example, coexist with many of the pro-urban myths highlighted above (something that was made apparent in our exploration of geographies of urban fear in Chapter 5). In such anti-urban myths the city is associated with sin and immorality, with a movement away from 'traditional' order and shared values. Biblical references to Sodom and Gomorrah, destroyed by God for sins against religious law, demonstrate the antiquity of this myth, while more recently the sheer ugliness of Victorian industrial cities, and the 'urban decay' experienced within contemporary urban areas, have inspired many journalists, writers and artists. John Ruskin, the nineteenth-century essayist and anti-industrialist, for example, said that cities are 'loathsome centres of fornication and covetousness – the smoke of sin going up into the face of heaven like the furnace of Sodom' (Ruskin, 1880, quoted in Short, 1991, 45). Here, the anonymity of the city, instead of providing an opportunity to be whatever one wishes to be, becomes associated with a sense of alienation, of not belonging (which takes us back to Relph's understandings of the modern human subject's detachment from place and Tuan's analysis of fear in the city: see Chapter 5).

You might notice here that some of the attributes of this anti-urban mythology also appear in pro-urban myths; the loss of 'traditional' values, for example, can signal either an irretrievable breakdown of social order (as part of an anti-urban mythology) or a liberation from oppression (as part of a pro-urban mythology). These mythologies are complicated creatures, then, shifting as they are explored from varying perspectives. Similarly, both pro- and anti-urban myths have worked themselves out in 'real' social situations in a variety of ways, impacting on the relationships between people and places in urban societies. We can illustrate this by examining the ideas that have encouraged some individuals and groups to reject urban living in favour of, for example, a 'back to the land' ideology or the idea of living in a 'Garden City'.

While it may seem odd to suggest that the atrocities perpetuated in Cambodia in the 1970s were caused by an anti-urban mythology, in some ways this idea helps us to make sense of what happened there. The Cambodian revolution was characterized by a forced exodus of the urban population from the cities to the surrounding countryside, where people were forced to engage in manual labour and peasant-style agricultural production, under an ideological regime which combined a communist (Maoist) revolutionary strand with an 'armed backwoods peasantry bent on destroying the degenerate civilisation of the cities' (Hobsbawm, 1994, 451). The urban middle classes were regarded as ideologically corrupt and were frequently tortured and killed, while many other former urban-dwellers, unskilled in farming, died from starvation, unable to produce sufficient food for themselves. The full horror of this is depicted in the film *The Killing Fields* (directed by Roland Joffé, 1984), and it is estimated that about 20 per cent of the population died under the revolutionary regime led by the infamous Pol Pot (Hobsbawm, 1994). The strongly anti-urban streak to this horrific situation, with its valorization of the peasant lifestyle, resonates uncomfortably, to some extent, with the 'back-to-the-land' movement of people such as Ruskin and William Booth in late nineteenth-century Britain. They advocated the simplicity and 'honesty' of small-scale agricultural production, along with engagement in arts and crafts, as an alternative to urbanized industrialism (Marsh, 1982). This led to the establishment of a number of experimental self-sufficient communities which paved the way for later 'green' communes. This illustrates that the way in which urbanism is imagined can take similar forms in very different times and spaces, yet have very different effects on the experiences of particular people in particular places and times.

While the Cambodian revolution and the late Victorian 'back-to-the-land' movement in very different ways illustrate anti-urban mythologies, the case of Garden Cities is rather more complex, combining pro- and anti- mythologies of both town and country. Garden Cities were the utopian dream of Ebenezer Howard, as outlined in his book *Garden Cities of Tomorrow* (1902). They were primarily a response to the rapid growth of London in the 1880s and 1890s, and particularly the development of slums and pollution associated with industry and its working-class population. Howard, with many other social reformers, associated the city with vice, dirt and squalor, at the same time believing in the potential of the city to produce a harmonious social order and adequate employment. The countryside, on the other hand, was associated with fresh air and beauty, while it suffered from being dull and narrow-minded. Howard's aim, therefore, was to design a new social order which combined the positive aspects of both the urban and the rural. As he put it, 'town and country *must be married* . . . out of this joyous union will spring a new hope, a new life, and new civilisation' (Howard, 1898, cited in Hall and Ward, 1998, 19). The result of this was Howard's famous plan for a 'Social City', itself a cluster of Garden Cities (Figure 6.1).

What you should note particularly in this diagram is its implicit **order**; the Garden City is *regular*, rather than chaotic, and it is *planned*, rather than organic. It thus draws on a specifically 'Modern' urban myth which values order and control (an idea that should lead us to question *whose* sense of order was being promoted here). Existing cities, apparently sprawling into the countryside, were perceived as threatening the neat demarcation of town and country – a distinction believed desirable at the time by many planners and architects (see Matless, 1993). Howard, as a member of the Victorian middle classes, had an interest in the reproduction of a particular social order, in the face of what must have seemed chaotic, uncontrollable and perhaps frightening urban expansion. This combined with his philanthropic motives to produce his Garden City, which was as much a vehicle of social engineering as a piece of urban design. There was a very specific geographical vision at work here, evident in Howard's assumptions about what the city should look like,

Figure 6.1 Planning the urban environment: Ebenezer Howard's 'Social City'

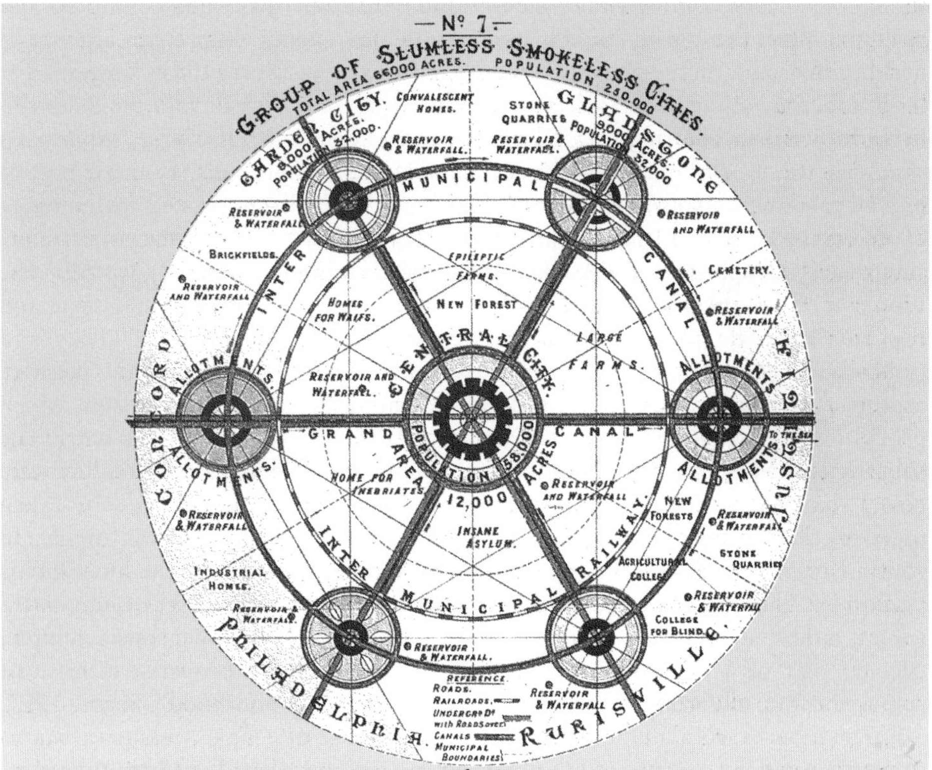

Source: Howard (1898)

which people should be where, and how they should live. In particular, the working classes seemed to be regarded as a 'problem', people who needed to be planned for and 'placed'. As such, the Garden City idea, while ostensibly challenging what was happening to cities during Howard's lifetime, maintained an established order of capitalism, property-ownership and middle-class morality. Because of this, perhaps, Howard's ideas became very influential in the subsequent development of town and country planning – several towns loosely modelled on his plans were developed from the early twentieth century onwards, including Letchworth and Welwyn Garden City (both UK), Canberra (Australia) and Siemenstadt (Germany).

Through the work of Howard and other planners, urban myths have come to shape the lives of people living in towns throughout the Western world. Indeed, in the twentieth century, a key idea driving town and country planning has been that the unrestricted growth of suburbia is a threat requiring the skilled and authoritative input of planners and cartographers in order to curtail and control its expansion (Hall, 1988). But from within the city, the view is somewhat different; suburban spaces have frequently been imagined as offering the potential for escape from inner-city 'intensity'. Like the Garden City, suburbia developed as a reaction to the squalor and chaos of inner-city residential and industrial areas. Slum conditions and pollution drove those who could afford it to the margins of cities, in an attempt to find healthier and more spacious lifestyles. However, as Short (1991) suggests, the term 'suburbia' also conveys a series of myths, signifying a break with the inner-city living rather than referring to a specifically delimited space. Of course, suburbia is characterized by differentiation along lines of class, wealth and ethnicity, but, as part of urban mythology, the (abstract) suburb is generally associated with a range of stereotyped lifestyles. Significantly, as neither the inner city nor the countryside, suburbia is imagined as encapsulating the best of both (imagined) worlds. The problems of the city are left behind, while urban services, facilities, employment and education are never far away. Space and fresh air are available, but rural 'idiocy' and isolation are absent. Suburbia is, then, the result of a drive to escape the negative aspects of urban life, without withdrawing wholly from an urban system which provides work, money and material goods. Pile (1999, 28) puts it this way: 'The suburbs were based on an assumption that communal life and orderly conduct were best fostered in neighbourhoods that were settled and domesticated . . . Taken together, the suburbs promised a calmer, safer, more prosperous way of life, in contrast to life in the city.' In terms of values and morality, where the inner city is licentious, 'The suburbs, in contrast, are used to refer to a whole set of alternative values: family, stability, security, a place where people settle down, raise children, become part of a community . . . . It gave a portion of property, a sense of achievement, a material and symbolic stake in the neighbourhood' (Short, 1991, 50). The suburbs were thus aspired to by a broad range of people, seeking the status a suburban residence conferred (a point emphasized in Osbert Lancaster's cartoons – see Figure 6.2). At the same time, suburbia was perceived to be an appropriate place in which to raise morally upright children (Aitken, 1998).

Figure 6.2 The suburban dream? 'Stockbroker's Tudor', Osbert Lancaster

There are, though, some important rejoinders to this positive myth of the suburbs. For example, it is apparent that the development of suburban areas contributed to a geographical separation of home and work (the private and the public) at a time when it was usual for men to go out to do paid work and women to undertake domestic work. We have already touched on this in Chapter 5, but in relation to this discussion of suburbia, men have tended to leave the suburbs to commute to the city centre on a daily basis, leaving women doing the housework and child-rearing in their suburban homes. One result of this has been a tendency to regard the suburbs as both domestic spaces and feminine spaces, and indeed to conflate the two, thus reproducing an association of women with unpaid domestic labour (Gregson and Lowe, 1995). For some women, then, the suburban dream can be restrictive in practice, although for others it can be seen as empowering, allowing the formation of new social and economic networks through participation in activities like Tupperware (or Anne Summers) parties, the Women's Institute or even the daily school run (Pile, 1999).

The second point is that suburbia has frequently been characterized as monotonous and dreary, composed of very similar houses, decorated and kept tidy in ways which conform to what have become expected standards. This part of the suburban myth refers rather mockingly to the cars which are washed every Sunday morning, the lawns trimmed until they resemble billiard tables, and the net curtains hung to preserve the privacy of those within and twitched nervously at any sign of anything vaguely out of place. For reasons of economy of scale, the suburban landscape is often characterized by a uniformity of building style (Chaney, 1994). One result of this is that relatively small differences in detail become much more significant than they otherwise would be, in terms of expressing status and individuality. However, because of pressure to conform to suburban norms, there is a limited range of socially acceptable ways of individualizing one's property (Chaney, 1994). Status and individuality might thus be expressed by having a particular model of car, or by putting plastic gnomes by the fishpond, whereas playing loud rock music or painting the house a violent pink are unlikely to be welcomed by the neighbours.

There is, then, a mythological set of moral standards which are important to suburban life – their inhabitants are imagined as heterosexual, monogamous and sexually unadventurous. The ironic subtext of these high moral standards is the equally mythical presumption of secret sexual liaisons, 'wife-swapping' parties, bearing the milkman's children, etc., which arise from a supposed situation of sexual repression. As such, we must note a long tradition of rebellion against suburban conformity and drabness, emerging in particular styles of music, such as Mod in the 1960s and the 1990s 'Britpop' of Pulp, Suede and Blur, as well as numerous 'situation comedies' which more gently parody the mores of suburbia. Examples include the sitcoms '2.4 Children', 'Terry and June' and 'Keeping up Appearances'; the last perhaps captures this best, focusing on the matriarchal figure of Hyacinth Bucket (pronounced, she insists, 'Bouquet'). Similarly, 'The Good Life', dating from the 1970s, followed a suburban couple who abandoned their life of safe conformity for an attempt to be self-sufficient in the back garden of their semi-detached house (much to the bemusement of their neighbours). Of course, this sitcom – like the others mentioned above – was given resonance by being set in a stereotypical suburban landscape (in this case Surbiton, in London's commuter-land). On other occasions, these landscapes have come in for more obvious vitriol, as in the poem by John Betjeman (published in 1937, two years before the outbreak of the Second World War) which begins:

> Come, friendly bombs, and fall on Slough
> It isn't fit for humans now.

Against this, the contemporary inner city has been viewed almost wholly through the lens of anti-urban mythologies created and sustained by the suburban middle classes. According to Short (1991, 50) 'the inner city is the dark underside of the city, a place of crime and disorder . . . . In the nineteenth century it was a place of the working classes, the source of disease and crime, the home of the crowd. In the twentieth it has become the locale of the underclass, a black hole of contemporary civilisation.' Nevertheless, certain aspects of the inner city, because of their relative exoticism, continue to hold a strange fascination for white middle-class people. For instance, the areas of inner cities often referred to as 'red-light districts' exemplify the way in which certain places exist in the collective imagination as spaces of both desire and disgust (see Figure 6.3). Places associated with prostitution are imagined as spaces of sexual immorality, risk and disease, but this association gives them a

Figure 6.3 Desire and disgust on the streets: a street prostitute in the UK

Photo: Authors

certain illicit attraction for those whose lives are lived out in more sedate and polite neighbourhoods (Hubbard, 1999). This blurring of the mythologies of the city, making ambiguous the distinction between simply 'pro-' and 'anti-', perhaps better reflects the true complexity of the social imagination of urban places.

This contrast between two specific parts of the city – the (abstract) suburb and the (abstract) inner city – demonstrates that the 'umbrella' urban myth can be fragmented into a multiplicity of smaller parts. Hopefully, this indicates that the mythologies of urban areas are complex and fragmented, but do nevertheless have significant effects on the way in which people live their lives in specific places. Inevitably, people experience and imagine 'real' places in relation to socially held myths or narratives at the same time that these urban myths impact on how particular places are planned and built.

■ Reading

Clearly, Short's (1991) book *Imagined Country* is a key reference for this section, and contains many additional examples of the expression of urban myths (among others). Peter Hall and Colin Ward's (1998) *Sociable Cities: The Legacy of Ebenezer Howard* is an interesting read, examining the cumulative impacts of Howard's ideas on urban planning in the twentieth century. We might also recommend Hall (1988), *Cities of Tomorrow*, as a highly regarded account of the 'intellectual history' of town and country planning – this shows how particular myths of the city have influenced specific architects and planners over the last 200 years. Chapter 4 – 'Consumer culture and suburban lifestyle' – in Chaney (1994) gives further detail on imagining suburbia, as do the essays in Roger Silverstone's (1997) *Visions of Suburbia*. Finally, Gerry Mooney's (1999) chapter 'Urban disorders' in Pile, Brook and Mooney (eds), *Unruly Cities: Order/Disorder* discusses processes of imaging urban disorder in contemporary society, working through a number of interesting case-studies.

■ 6.4. Wild and natural places

Wilderness is often imagined as the opposite of civilization. Yet if, as implied above, civilization (or urbanism) is a highly problematic concept, then understandings of wilderness and nature must be equally complex. For many, nature is something which is 'real'; it is 'out there' in the form of plants, animals, rocks, weather, water, etc., and the relationships between those elements. This 'scientific' understanding opposes 'nature' and 'culture' (that is, human systems of meaning, value, etc.) in rather the same way that we often think about the divide between urban and rural. Contemporary human geographers, however, have tended to understand nature in a very different way, exploring the ways in which 'nature' is **socially constructed** and imaginatively brought into existence by particular social groups (e.g. Whatmore, 1999). From this perspective, nature, far from being an unchanging set of physical realities, exists as a mythological set of ideas, in the much the same way as the (abstract) urban does. As with urban myths, this does not mean that our ideas of nature and wilderness are unimportant. Instead, the ways in which we construct (collectively think about and understand) nature have important impacts on the way in which we live in and engage with the physical environment, other (plant and animal) species and other human beings. In this section, we hope to sketch out some of the most important dimensions of this, although space is insufficient to explore them fully: you should consult the suggested reading at the end of this section for more detail. Beginning with the concept of 'wilderness', this section will look at anti- and pro-wilderness myths (again drawing on Short, 1991), before looking at what these myths say about our relationships with the physical environment and other people.

The concept of 'wilderness', according to Short (1991), originated from the first agricultural revolution around 10,000 years ago, when the first groups of people moved from a nomadic hunter–gatherer lifestyle to a sedentary farming lifestyle. In hunter–gatherer societies, there can be no such thing as wilderness; the whole environment is used to supply food and other materials, and thus there is no differentiation of the land surface. As soon as farming begins, however, some land becomes farmland, and is hence immediately distinguishable from non-farmland. The concept of wilderness, then, relies on a distinction between farmland and wilderness. In this opposition, the farmland is associated with being settled, with being cultivated, with domestic animals, with relative security and stability. In sum, settled farmland is imagined as 'central' to people's lives. On the margins, away from the settled centre, the wilderness, as an 'other', is associated with wild animals, wild 'uncivilized' people (outlaws or savages), and with danger and insecurity. From this arise important sets of myths or social imaginations about wilderness, which can be loosely characterized as anti- and pro-wilderness. As with pro- and anti-urban myths, these mythologies necessarily exist simultaneously, but it may be suggested that, very broadly, the anti-wilderness myth was dominant until a couple of hundred years ago, and that since then, a more Romantic pro-wilderness myth has increased in significance.

Viewed from the anti-wilderness perspective, wilderness may be considered as that which is outside civilization. Indeed, this exclusion of wilderness is an important part of civilization. The transformation of wilderness and nature into farmland (and later, urban-industrial society) is regarded as a triumph of social progress, a conquest of a brutal 'real' nature. Wilderness itself, in this myth, is something or somewhere to be conquered or feared. The desire to conquer, control and command nature is illustrated in a range of human activities, from scientific experimentation, through to air-conditioning and genetic modification of crops, and from mountain-climbing to off-road driving and 'extreme sports' (see Cloke and Perkins' (1998) article on adventure tourism in New Zealand). This feeling that wilderness is the opposite of civilized control over nature is, for example, important for Britain's National Farmers' Union as they seek additional financial support at a time of great difficulty for many farmers; the threat of the farmed countryside returning to wilderness is one that seemingly prompts political concern (*Guardian*, 1 September 1999). Similarly, fear of the wilderness is clearly expressed in folklore and children's stories where impenetrable forests are feared as places to get lost in (as in *Hansel and Gretel*), or where one will encounter dangerous animals (like the wolf in *Little Red Riding Hood*). Going into these places represents a fear of leaving home, security and civilization. More subtly, they may also be concerned with an implicit understanding that our own assumed 'civilization' is little more than a thin veneer, covering a 'wilder', more 'natural' set of desires which are just waiting to be expressed. In this vein, it has been suggested that, for example, *Little Red Riding Hood* can be read as a young girl's growing awareness of her (potentially dangerous) sexuality, given physical form in the body of the wolf (see Bettelheim, 1976). Similarly, this fear of the 'repressed' wilderness is expressed in more 'adult' literary works (though perhaps fairy-tales like *Little Red Riding Hood* have origins as adult stories as well, allowing adults to express and deal with social fears as part of a process of enculturing children). Going back to Jane Austen's *Pride and Prejudice*, the fragile boundary between what is imagined as 'civilized' and 'uncivilized' behaviour is noted in the following exchange between the pompous Sir William Lucas and Mr Darcy:

> 'What a charming amusement for young people this is, Mr Darcy! There is nothing like dancing, after all. I consider it as one of the first refinements of polished societies.'
>
> 'Certainly, sir; and it has the advantage also of being in vogue amongst the less polished societies of the world – every savage can dance.'

In a somewhat similar fashion, Shakespeare uses the concept of wilderness in some of his plays to symbolize a breaking down of the civilized and ordered. The witches in *Macbeth*, for example, are evil and frightening creatures encountered in the wilderness, and associated with wild weather and crimes against civilization and social order (the killing of a king). When the king is killed, domestic animals become wild (horses eat each other) and people are seen to lose their humanity, in a dramatic breaking down and reversal of the imagined boundaries society has created between 'wilderness' and 'civilization'. In *King Lear*, the eponymous king's growing madness is symbolized by his return to a state of wild(er)ness – he tears off his clothes and leaves his home to go out into the wild during violent storms, while his kingdom topples around him. The very word used to describe Lear's madness (he is 'bewildered') contains rooted within it the 'wild' that he has become. There is a direct association here between madness and wilderness; both contrast with the expectations of civilized, rational, settled behaviour; the normal is opposed to the abnormal, civilization is opposed to wilderness. Lear is likened to the turbulent ocean and associated with a list of weeds, plants which smother the cultivated crops of settled farmland. He is

> As mad as the vexed sea, singing aloud,
> Crowned with rank fumiter and furrow weeds,
> With hardokes, hemlock, nettles, cuckoo flowers,
> Darnel, and all the idle weeds that grow
> In our sustaining corn.
>
> (*King Lear*, Act IV, scene iv, ll. 2–6)

What is indicated by the above is that social understandings of wilderness are as much, if not more, to do with understandings of *ourselves*; they are ways of thinking about our social and individual identity, as supposedly civilized beings.

On the other hand, the *pro*-wilderness myth begins to question this perspective of wilderness as negative and civilization as positive. This dimension of the wilderness myth rather reverses the picture, in that it recognizes the negative side of civilization, and develops alternative, romanticized, understandings of wilderness. As we saw above, the increasing urbanization and industrialization of Western European society during the eighteenth and nineteenth centuries was associated with the development of specific anti-urban myths, and concurrent with this was a new-found reverence for wilderness and 'unspoiled' nature. So where pro-urban myths can be opposed to anti-wilderness myths, the recognition of negative aspects of urban society (poverty, crime, exploitation, material greed, the ugliness of urban-industrial sprawl) can be opposed to a simultaneous construction of the 'good' in nature. Particularly among the upper classes, it became fashionable to dissociate oneself from industrial production and material possession (after all, with inherited wealth, one need not concern oneself with vulgar industry – see Chapter 7). Instead it was deemed appropriate to promulgate a romanticized myth of a nature with which one could have a deep spiritual relationship, in contrast to the shallowness of the urban veneer. Wild places underwent transformation, then, but this time not materially, in order to become of useful service in agriculture or industry, but imaginatively and spiritually, as places where one could be liberated from social convention and manners. An aesthetic transformation also occurred, so that the wilderness which could once be described as 'full of horrors: dreadful fells, hideous wastes, horrid waterfalls, terrible rocks and ghastly precipices' (Thomas, 1983, 258–9), became regarded as visually pleasing and spiritually uplifting places to be. As Evans (1997, 21) puts it, 'mountains and moorland gradually acquired a reverence for their awesome and brooding beauty'.

These ideas clearly affect contemporary Western understandings of leisure and recreation, with the mountain areas of Britain's National Parks, for example, ironically coming under intense pressure from the hordes of visitors wishing to experience the solitude and beauty of a 'wilderness' landscape (see Chapter 1). As an example of this different imagining of the wilderness, consider this excerpt from Samuel Taylor Coleridge's 'Frost at Midnight', a gentle and moving poem written by the eighteenth-century Romantic poet to his baby sleeping at his side:

. . . For I was reared
In the great city, pent 'mid cloisters dim,
And saw naught lovely but the sky and stars.
But thou my babe! shalt wander like a breeze
By lakes and sandy shores, beneath the crags
of ancient mountains, and beneath the clouds,
Which image in their bulk both lakes and shores
And mountain crags: so shalt thou see and hear
The lovely shapes and sounds . . .

In this poem, civilization's buildings and dark interiors are represented as a prison ('pent 'mid cloisters dim') while the wilderness, with its freedom and beauty, is celebrated as the ideal place for a child to roam (and compare this with Little Red Riding Hood, for whom the wilderness signifies danger). Around the same time, Wordsworth's pæans to the Lake District performed a similar role in fostering the idea that rural life had a simple and rugged beauty; 'Michael', the story of a shepherd's son, famously suggested that wild hills imparted 'a stout heart and strong head'. Nevertheless, it is incorrect to say that Western societies made a simple transition from a fear of wilderness to a reverence for wilderness. Underlying an appreciation for 'natural beauty' and solitude is often a sense of unease in places far away from 'civilization'. Films such as *The Blair Witch Project* (1999, directed by Eduardo Sanchez and Dan Myrick) can exploit this fear of the unknown wilderness to terrifying effect; the film's reliance on psychological manipulation rather than explicit special effects makes it particularly horrifying, while illustrating that our fears are culturally and psychologically generated, rather than being innate to a specific type of place or landscape.

A further dimension to this consideration of wilderness places is the way in whichthe people living in the wilderness are imagined. As with the imagination of the physical environment and landscape of wilderness, this can be demonstrated to have changed over time, with important implications for the ways in which 'civilization' has interacted with those people, particularly during periods of imperialism. In particular, people living in the (abstract) wilderness have been regarded as a social 'other' to those living in 'civilized' places, invested with characteristics that identify them as 'uncivilized'. It is essential to understand here that what is under discussion is less the 'real' characteristics of any one group of people and the place they live in, and more the way in which those groups are imagined. As we have seen, myths tend to have pro- and anti-dimensions, and in mythologizing wilderness people, 'civilized' people have been able to invest them with good and bad characteristics, invoking the good and bad aspects of 'civilization' we have already mentioned. And this is the key point here. Myths held by 'civilized' people about 'wilderness' people can say more about the identities and preoccupations of the 'civilized' than indicating the 'true' characteristics of 'wild' people and places, with the 'uncivilized' acting as a necessary 'other' or opposite, to civilized identity. In this sense, those people and places that are imagined to exist on the margins of civilized society are used to define what that civilization might actually be. The fact that there are both good and bad sides to this reflects the complexity and ambiguity of understanding one's own, 'civilized', identity.

Representations of 'uncivilized' people, therefore, have had two main dimensions. Firstly, there is a long history of understanding them as 'savage', and particularly from a Western imperialist tradition, justifying colonialism as the control of wild and savage people by the civilizing forces of European culture. Racist stereotyping of black people as lazy, ignorant and uncontrollable is related to this process of justifying white imperial expansion into the 'heart of darkness'. For example, Peter Jackson (1989, 135) illustrates this when he cites these racist words from the English *Gentleman's Magazine* of 1788; 'The Negro is possessed of passions not only strong but ungovernable . . . a disposition indolent, selfish and deceitful . . . . As to all the other fine feelings of the soul, The Negro, as far as I have been unable to perceive, is nearly deprived of them.' Such stereotypes were reinforced by a Christian morality which encouraged the conversion and control of non-Christian 'heathens'. A fear of the uncivilized 'other' was reproduced by representations of cannibalism and (non-Christian) religious practices associated with uncivilized people in wild places. Tales of European heroism in the face of savagery heightened these myths while reproducing racist stereotypes. Referring to Figure 6.4, Jackson (1989, 138) comments 'Comics like *Boy's Own* were extremely popular reading for the urban middle classes, depicting the bravery of the solitary white explorers overseas, confronting the "yelling masses" of irrational, violent, angry natives.' Similarly, Gallaher (1997) writes that white American identities required the presence of a series of 'others' – characterized as the gothic, the dark and the evil.

More recently, these negative representations of 'uncivilized' people have been balanced by more sympathetic, though ultimately equally stereotypical representations of those who are 'other' to white people in Western urban societies. Along with the emerging pro-wilderness sentiment, alternative myths of wilderness people emerged in the midst of the Romantic backlash against urban and industrial lifestyles. Instead of being feared, these people were regarded as possessed of a certain nobility and spirituality (despite their lack of manners and 'civilized' culture) which came from a close relationship with nature. Hence the figure of 'the noble savage' (associated with the eighteenth-century French Romantic Jean Jaques Rousseau) is important in the way that urban societies have imagined their 'other'. For Rousseau, what was commonly regarded as progress was in fact a way of binding people in chains (of work and social convention): 'Rousseau maintained an attack on civilisation and praised the idea of man [*sic*] close to nature, free from constraints of contemporary society and morality. . . . For the romantics, hunting–gathering groups became a vehicle for criticising their own societies' (Short, 1991, 22–3). Yet again, instead of these representations necessarily revealing some truth about 'uncivilized' people, they are largely a

Figure 6.4 Racial stereotypes: the 'civilized' and the 'savage'

Source: Jackson (1989)

reaction to something Western people see in their own identities; that is, the feeling that urbanism and 'civilization' has corrupted 'pure' experience of the world. The supposed depth and spirituality of 'the noble savage' therefore becomes an attractive myth.

Ultimately, we have to recognize the close similarities between social mythologies of (urban) civilization and wilderness. They are both, despite or because of their own complexities and ambiguities, about what it means to be human and to live within particular sets of social and environmental relations dictated by particular social groups. These ambiguities are illustrated in the recurrence of particular ideas as facets of both sets of myths, so that in different ways, both order and disorder can be seen as parts of both civilization and nature. For example, 'civilization' can be imagined as *ordered* in a social hierarchy based on social class, but *disordered* in relation to crime. 'Nature' can be imagined as a set of *ordered* scientific laws and biological rules, and *disordered* in its disregard for social manners. There is perhaps in all this a human desire for order and knowability, but an intense recognition that behind any impression of order lurks an unknowable confusion which threatens our sense of stability, right and wrong, self and other, and place in the world. In this sense, contemporary concerns over the instability of society–nature boundaries are played out in the way that some 'wild' animals (urban rats and foxes, for instance) are feared and mythologized for disturbing the normally heavily policed boundaries between urban and non-urban spaces. Such animals have been regarded as 'fissures' in the fabric of the contemporary city, allowing nature to emerge into inappropriate places (Davis, 1998). The rightful place for nature in cities, it is felt, is either in zoos (where a controlled and safe experience of the exotic can be had, often in the context of scientific education about 'nature' – see Anderson, 1995; 1998) or in the form of pets (animals which are frequently regarded by their 'owners' as semi-human and socialized). This ambiguity over categorical boundaries, where, on the one hand, a pet dog may be regarded as semi-human and, on the other hand, certain groups of children have been portrayed as animal, feral or vermin (as in myths of slum-children living in sewers), illustrates the fragility of the categories 'society' and 'nature', 'civilized' and 'wild'.

■ Reading

Again, Short's (1991) *Imagined Country* contains much relevant material in the chapter on 'Wilderness'. David Pepper's (1996) excellent book, *Modern Environmentalism: An Introduction*, tells the story of the changing relationships between society and nature, detailing the social construction of nature and wilderness. The first chapter of philosopher Kate Soper's (1995) book *What is Nature? Culture, Politics and the Non-human* is also worth reading if you find these ideas interesting. Wolch and Emel's (1998) edited volume *Animal Geographies* explores an increasingly important dimension of this issue which has not been covered here (the best parts to read in this context include the preface and introduction, Chapter 4 '*Le pratique sauvage*: race, place and the human–animal divide', and Chapter 5, 'Are you man enough, big and bad enough? Wolf eradication in the US'). Elsewhere, Kay Anderson (1998) and Whatmore and Thorne (1998) examine the social construction of 'wild' animals in various human contexts (e.g. zoos and gladiatorial combat). Discussion of race, racism and the representation of 'uncivilized' people is explored in Chapter 6 – 'Languages of Racism' – of Peter Jackson's (1989) *Maps of Meaning*. This chapter also touches on issues relating to Orientalism (examined below) and is particularly recommended for that reason.

6.5 The mystical East: imagining elsewhere

In this final section of this chapter, some of the ideas emerging from the discussions of geographic myths will be applied to the concept of 'the Orient'. The construction of an imagined, mythical Orient, existing on the (geographical and social) margins of Western 'civilization', has been an important way in which countries of the West have justified colonialism and tried to understand their own identities in relation to an imagined East. This is particularly the case for nations (such as Britain and France) which have had extensive colonial involvement in the East, with Said (1978, 1) arguing that 'The Orient was almost a European invention, and had been since antiquity a place of romance, exotic beings, haunting memories and landscapes, and remarkable experiences.' The rest of this section develops this idea, with particular reference to the work of Edward Said (a Palestinian living in the USA), whose book *Orientalism* (1978) has had a profound influence on the ways in which geographers have understood Western society's relationship with the East. By 'the East', Said refers particularly to the 'Near East' and North Africa, but his ideas have been extended to include the Middle and Far East. Said attempted to show that Western attitudes to the East were racist (Orientalist), in that they relied on stereotyping of Oriental people and cultures in order to support a feeling of Western superiority. In the same way as we have suggested pro- and anti- urban and wilderness myths, positive and negative dimensions to Western imaginings of the Orient as marginal to the West's centrality can be demonstrated. The positive dimension, for example, allows people in the West to imagine the Orient as a sort of exotic paradise, a place of luxury, leisure, sexual pleasure (the myth of the harem), seductive perfumes and mysterious magic (the flying carpet and snake-charmer stereotype). Oriental people are characterized as charmingly hospitable, while attention is drawn to Eastern philosophies which emphasize the spiritual and physical harmony, such as in the martial arts or Feng Shui. The corresponding negative dimension of the Orientalist myth associates Eastern cultures with decadence, idleness, barbarism and lack of civilization. Miles Ogborn (1988), for example, writes about the Victorian scandal sheets (equivalent to tabloid newspapers) reporting on the barbarism and sexual backwardness of the Orient as compared to British civilization. In this negative myth, Eastern people are regarded as callous, inscrutable and cruel, and it is this barbarism that undermines their civilized status.

Such myths have been regularly played out in Western society, in the accounts of explorers such as Sir Richard Burton (Figure 6.5), in films and novels (such as *The Phoenix and the Carpet* and *Arabian Nights*), and computer games involving martial arts. In descriptions of Oriental cities, this trend is particularly prevalent. The Oriental city is characterized as teeming, chaotic and convoluted, its streets littered with 'filth, food, beauty and blood' (Jervis, 1999, 89), a city undermined by decadence and immorality. Advertising also makes use of Orientialist myths. A recent UK advert for Peugeot cars, for example, used stereotypical images of the desert, camels and desert people, along with music and ideas of luxury, mystery and sexuality. The advert showed two Western women stuck in the desert in their Peugeot, and being rescued by a stereotypical Arab Sheikh, who invited them back to his home, which contained a harem of wives. The joke is that the two women assume that the hospitable yet vaguely sinister figure of the Sheikh wants them to join his harem, while in fact he actually wants their car, and eventually they make their escape by speeding off into the desert. This advert encapsulates the Orientalist myth, being a mixture of the sensual and exotic with a sense of fear and entrapment, couched in a humorous misunderstanding. It contrasts the apparent freedom and independence of the Western women in their Western car with the imagined subservience of the women in the harem.

Clearly, as with the social imagination of the wilderness, these positive and negative myths of the Orient should not be taken to be accurate representations of 'real' people and places; indeed they can be highly offensive – the myth of Oriental sexuality, for example, may be offensive to those Muslims who see Western capitalism and sexual promiscuity as corrupting influences. Instead, they can be regarded as a way in which Western identities have been formulated in relation to an imagined 'other'. Thus,

> the Orient was seen in terms of an implicit contrast with the West. This process is not a calm and logical one, however; it is full of emotion, fantasies, fears and desires. The Orient was all that the West both feared and wanted in itself; hence the complexity of its simultaneous praise for the spirituality of the desert people and its dislike of what it saw as Oriental customs.
>
> (Rose, 1995, 104)

Recognizing the positive and negative dimensions of Orientalist myths allows us to understand something of the complexity of identity. Importantly, because there are both 'good' and 'bad' facets of the imagined Oriental 'other', there is a reciprocal understanding of the desirable and undesirable qualities of Western society. More subtly, perhaps, what is presented as undesirable is, paradoxically, actually often highly desirable; hence, the sexual decadence associated with the Orient is imagined as bad (as it supposedly leads to idleness and a breakdown of social order), but it is

Figure 6.5 The Oriental myth: Richard Burton's 'Personal Narrative' (1893)

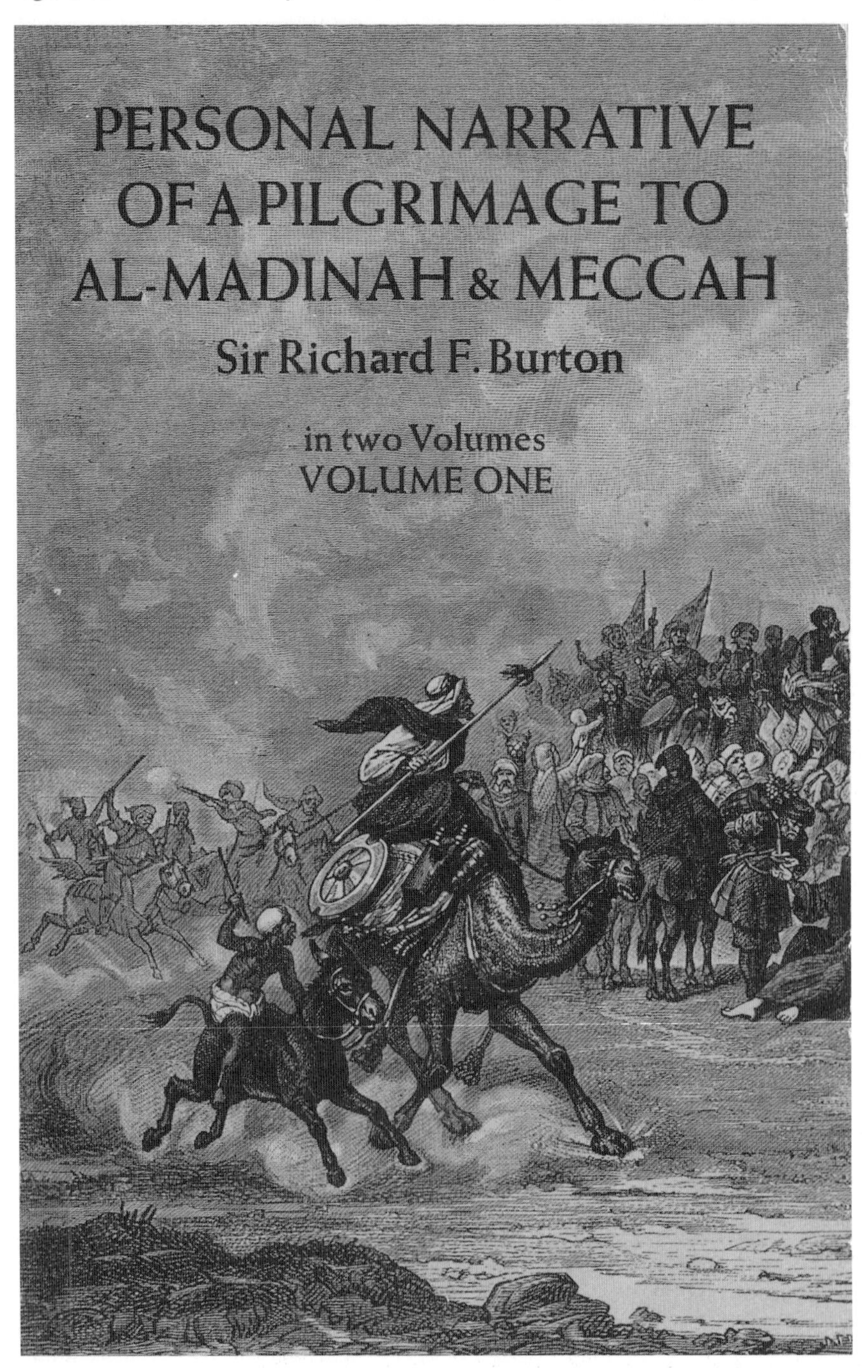

Source: Burton (1964)

also alluring and desirable (hence the fascination with such things as the *Kama Sutra*). Orientalism needs to be seen as a power-laden myth in so much that it involves racist stereotyping associated with the idea of Western superiority, and gender stereotyping in its representation of Eastern women as exotically, mysteriously and passively sexual. Beyond its function of being the necessary 'other' to Western identity, Orientalism has had important effects on Western relations with the East. We can cite two examples (mentioned in Rose, 1995) to show this. Firstly, the representation in many paintings and written reports of the crumbling remains of Ancient Greek and Roman settlements in the East fed Western assumptions that Oriental people were unable to look after their heritage 'properly', and that important artefacts should be brought 'back' to the West in order to be 'protected'. Secondly, there is clear evidence of continuing Western perception of Islamic fundamentalism as a threat to Western society, influencing the USA's (along with other Western countries) involvement in conflicts such as the Gulf War (1991). Imagined or mythical understandings of 'other' people and places, then, have real impacts on the relationships between different societies (Rose, 1995).

As a final comment, recent manifestations of Orientalism have emerged in relation to countries like Japan, which has developed into a powerful industrial and financial state since the Second World War. Here, the imagined inscrutability of Oriental people appears to have been combined with hard work and devotion to one's employer in order to produce an economic threat to Western countries. Fear of this economic power is coupled in the Western imagination with admiration for the cultivation of so-called 'Peoplism', but nevertheless both are combined in Orientalist myths which, while different to those which preceded them, recombine similar elements in new ways. We might therefore finish this section by asking you to think about the myths associated with a variety of 'Japanese' goods and products, ranging from sushi, noodles and sake through to Sony Walkmen, Kung Fu films and Suzuki motorbikes. Which of these are thought of as Western goods, and which as Oriental?

■ Reading

Gillian Rose's (1995) chapter 'Place and identity: a sense of place' contains a useful discussion of Orientalism. Edward Said's (1978) book *Orientalism* is the source of many of the ideas in this chapter; an excerpt from this can be found in Agnew, Livingstone and Rogers (eds) (1996) *Human Geography: An Essential Anthology*. Chapter 1 – 'Foreign places' – of *Worlds of Desire, Realms of Power* by Pamela Shurmer-Smith and Kevin Hannam (1994) is recommended as an exploration of the exoticism of the foreign in the geographical imagination, while May (1996) ties this into the 'everyday' by considering how we consume the exotic.

Summary

In this chapter we have begun to look at the importance of myths in shaping the way in which people think about particular people and places. Here, a number of ideas have been suggested as fundamental for helping us to understand how people and places come to have (spatially specific) identities:

- Myths are part of the way in which we define ourselves and the places we live in, often through opposition ('us' and 'not us'). Imaginative geographies ascribe meanings to people and places, often through the opposition of 'self' and 'other'.
- Place identities are often highly ambiguous so that the same characteristics can be regarded simultaneously as 'good' and 'bad' (and the same is true of identities of people).
- Although myths may not be accurate representations of 'real' people and places, they can and do have actual effects on the ways in which groups of people interact with other people and with their physical and social surroundings, playing a role in the history of planning, politics and internal relations – all of which impinge on the everyday.

seven Representing place

This chapter covers:

7.1 Introduction

For many geographers, an important way that we come to know about people and place is through their **representation.** Indeed, we have already touched on the significance of representation in this book in various ways. In the last chapter, for instance, we referred to numerous forms of representation (poems, song lyrics, films, paintings) which serve to show how places are imagined by different social groups. Similarly, in Chapter 2 we suggested that television, film, pop music and video are responsible for communicating images of diverse places to an ever-increasing global audience. Considering these representations, we might distinguish between those that involve a deliberate attempt to communicate something about a place (like an advert for a holiday resort or a news item about a specific part of the world) and those that use place as a 'backdrop' in order to exploit positive or negative associations of that place (for example, advertising which associates fresh food products with images of the countryside or a police drama that seeks to exude gritty realism by being set in the inner city). In both cases, the individual or group which creates the representation (i.e. its 'author') is using place imagery for a particular reason, hoping that people will respond positively or negatively to the images of place being communicated.

Yet geographers' interest in representation is somewhat broader than this, as all representations potentially communicate place myths in one way or another – even if that was not necessarily the intention of the author who created the representation. For instance, globally marketed CDs by acts such as Clannad, Sinead O'Connor or Enya may summon up various images of Irishness for those who buy their music (the majority of whom have never been to Ireland), while performers from other nations may be only dimly aware that their music represents a version of that nation to a wider audience. Genres like Country and Western may even come to symbolize a whole set of national values for some people (to the extent that many country fans around the world have adopted 'American' fashions and social rituals like line-dancing). Forms of music thus conjure up images of other parts of the world as particular types of instrumentation and composition are taken to be endemic to particular nations, regions or cities. Of course, these are often the result of **cultural hybridity** (see Chapter 2), as in the case of Merseybeat, the fusion of black rhythm and blues and white Country and Western music that emerged in Liverpool in the 1960s as bands like The Beatles and Gerry and the Pacemakers drew on the influences present in a busy port.

From pop music to poetry, film to fiction, it is obvious that representations can include a vast array of artefacts and forms which people use to interpret the world around them and present themselves to others. To make sense of this array, what we will try to do in this chapter is to consider a number of ways in which geographers have tried to make sense of representations. As we will begin to explore, it is never simply the case that geographers take representations to reflect what 'really' exists in the world. Instead, they tend to focus on the partial and selective nature of representations, exploring how this **selectivity** contributes to the production of place in people's imaginations by emphasizing some things and ignoring others. An important concept here is that this selectivity is as true for a seemingly 'objective' representation (a newspaper report or TV documentary) as it is for a 'subjective' representation like a song, poem or painting. Collectively, though, representations of a place are very much responsible for shaping the meanings which different individuals attach to place. In turn, these representations can affect people's experiences of place and their behaviour in place. Further, representation of place is often highly contentious, as there can be a range of different ways in which specific places are represented. Representation can thus be seen as part of a struggle to control places and the people in them (something we will expand upon in Chapters 8 and 9).

Before we consider these ideas, we perhaps need to begin with some more basic definitions. To start with, we are using the term 'representation' in this chapter in a very general sense, referring to any means of communication by which people tell each other about the world and share information. These take a variety of forms, including writing, speech, music, painting, fashion, computer graphics and 'body language'. In our everyday lives, we are constantly surrounded by representation, as people are always trying to tell us things about themselves, about other people, about home places and foreign places, about objects or about experiences and feelings. People may be wanting or expecting us to buy something from them, to agree with their opinions about particular people and places, to be persuaded to vote in a particular way, and so on. To do so, they must communicate their message to us in a form we understand, or else the message is lost. We need, therefore, to consider the processes that allow this communication to occur and explain how representations spread ideas about the type of people we might expect to encounter in particular places. To these ends, we think through questions of what representations are and how they 'work'. Two case-studies are explored in depth, to show that the ways in which we understand and use places are intertwined with the ways in which they are imagined and represented in, for example, novels, films and paintings. The first example concerns historical and contemporary representations of the countryside while the second concerns stereotypical representations of the 'North' and 'South' of England. We want to use these examples to stress that representations are highly problematic, being bound into the complex relationships (and sometimes antagonisms) that exist between different cultural groups. Finally, the ways in which geographers themselves are implicated in processes of representation will also be explored, with attention given to the way that our understandings of the Internet are understood geographically through the representation (and mapping) of 'cyberspace'.

7.2 Interpreting communication: what is representation?

Essentially, representation relies on *signs* to convey meaning. We need to interpret 'signs' very broadly here; clearly, written words are one type of sign, but colours, sounds, objects or textures can communicate meaning too. The study of the way that signs communicate meaning is generally referred to as **semiotics** or **semiology** (literally, the science of signs) after the pioneering work carried out by Ferdinand de Saussure and Charles Pierce in the early years of the twentieth century. On the most basic level, semiotics examines the relationship between signs, meanings and the individuals who are 'reading' those signs. In essence, every sign is taken to have a specific meaning; for example, a red light means 'stop'. However, the relationship between the sign and what it means (or signifies) is only fixed in relation to a specific audience: for instance, to some people and in some contexts, red lights may mean sex for sale. Extending this type of semiotic analysis, it becomes obvious that a wide variety of artefacts are capable of communicating meaning: painting, music, architecture, fashion, interior decorating and landscape gardening all have the capacity to 'speak' to different groups in different ways. When people choose clothes, for example, they do so not only on the basis of price, durability or fit but also on the basis of the messages they wish to communicate about themselves. Alison Lurie (1992) accordingly argues that clothes are a key means by which people choose to express their tastes, opinions and interests, with certain fashions taken as indicating the style and prestige of their wearers.

Representation, then, basically implies the existence of a shared **system of meaning** which people draw upon in a variety of ways in order to communicate. Of course, if some people do not understand or acknowledge this system of meaning then the message is lost on them. An obvious example is that of the spoken word, where the existence of different languages, dialects or slang terms may preclude effective communication between individuals. Yet this 'problem' holds for other types of sign – such as when the clothes you think are sending out a message that you are sexy, fashionable and stylish are interpreted as drab, old-fashioned and ugly by others. Shared systems of meaning imply more than shared language then: they imply shared ways of looking at the world, a shared outlook, a shared interpretative framework. Consider a further trivial example – a cheap, disposable, biro pen. As a meaningful object, discussing the pen requires first that you, the reader, recognize the word 'pen', and associate it with the object that we, the authors, intend you to. The word 'pen' is, however, simply an arbitrary and convenient label allowing us to communicate with you. We agree, implicitly, that the object referred to is a 'pen', but by doing so we have attached a meaning to the written word – you associate it with the object. In doing so, we are involved in creating a representation of the object. We allow the word 'pen' to represent, to stand for, a plastic thing that one writes with. If you were to read the word out aloud – 'pen' – you would be making a further representation – this time an audible representation of a written word. Other culturally specific meanings can take this understanding of representation further – the pen can become more than a pen. Our cheap pen can have meanings of, for example, disposability attached to it. If it has the name of a seaside resort – 'Blackpool' – inscribed on it, the pen can become a memento. If the pen is given by one person to another, it becomes a gift, with value beyond its financial cost – and this is particularly meaningful in a society where gift-giving strengthens interpersonal relationships.

It can be seen that the (by now very meaningful but still cheap) pen is associated with other, perhaps more important, sets of meanings and social relationships. When, for example, the biro is given by a wife to her husband as a gift, it becomes, in a small way, symbolic of their marriage. The pen, as a gift, becomes meaningful within the context of a specific social institution. In some ways, then, the pen (as a gift) can represent, or stand for, the marital relationship itself. In the event of the death of one of the partners, or simply their separation, this pen would take on different meanings, reminding one partner of the continuing absence of the other. Here, we have moved from thinking about a small plastic writing implement to the ways in which people live their lives, by considering the ways in which objects and interpersonal relationships become meaningful – at least to those people who broadly share the same systems of meaning. Of course, this leads us to the question as to where these meanings come from and what role people play in creating systems of meaning. Following the sociologist Stuart Hall (1997), we can suggest that the meanings of signs are more or less consistent across **cultural groups** – that is to say, groups of people who adopt similar ways of life and whose everyday lives are given shape through the cultural artefacts and rituals they create and exchange. As such, we need to note that even the meaning of marriage itself may be seen to differ across different cultures. After all, meanings are susceptible to change and reinterpretation by different groups. Marriage, for example, has been criticized as an institution which involves the subordination of women and marginalizes non-heterosexual couples, but this type of criticism cannot be levelled at recent gay weddings between gay men or lesbian women. Within the English language, then, words often take on different meanings for different cultural groups as well as changing meaning over time. As an example, if we say something is 'awful', we usually want to attach negative meanings to it. In the eighteenth century, however, the word implied that the speaker was 'full of awe' – literally amazed – at whatever was being referred to.

Hall (1997) stresses that there are three main ways of thinking about the way representations communicate meaning:

- The **reflective** or **mimetic**. This suggests that meaning lies within what one is representing. The process of representation is therefore seen as simply communicating the 'true' nature of whatever is being portrayed. By using signs, it is considered that it is possible to provide an accurate reflection, a mirror image, of what exists in the world (for example, a place). This view is limited, however, in so much as it doesn't recognize the ways in which the author brings something to the representation, rather than simply passing on what already exists. We have already criticized humanistic assumptions of authentic *genius loci* for very much the same reason (see Chapters 4 and 5). Hence the second way of understanding representation is . . .
- The **intentional**. This conceptualization of representation relies on authors being able to use language to communicate unproblematically what they want to say. Here, the words, pictures or sounds mean what their authors intend them to. Again, this is problematic because it doesn't account for the ways in which the representation is received by different cultural groups in different contexts. This leads to a third way of thinking about representation . . .
- The **constructivist**. This approach recognizes that meaning is something which is culturally constructed (i.e. bought into existence) in the midst of the varied interrelationships that occur between different people. Meaning is hence produced *by* communication (in speech, writing, film, body language . . .), rather than simply being transmitted *through* communication. The languages of speech, writing, painting (and so on) are constantly kept alive by people communicating with each other; there is a separation between on the one hand, the symbolic world of signs, and on the other, the material (real 'things', like tables and chairs) and immaterial ('ideas', like love and pain) worlds to which they refer.

From this latter perspective, representation is involved in our understanding of the material and conceptual worlds – including the *creation* of people and places in our imaginations. In turn, these representations of people and place – as good or bad, moral or immoral, desirable or disgusting – will affect the ways in which the places and people being represented are related to (Sibley, 1995). For example, media representation of inner-city areas as dangerous and threatening may discourage people from going there, and engender particular attitudes towards them from authorities like the police and town-planners. Equally, the representation of asylum-seekers attempting to enter the UK as 'scroungers' by certain sectors of the British media contributes to the shaping of public opinion and political debates about how such people should be treated. Often, such representations bear little relation to the 'reality' of a place or a group of people, but they 'speak' to particular cultural groups in a way that leads them to be accepted as a common-sense view of the world. Representation is far from innocent, then, being part of a complex cultural struggle for certain views and opinions to be accepted as normal and correct. Accepting that people and places are represented in particular ways for particular reasons, one of the principal tasks facing geographers is to try and examine how and why people and places are represented in such ways.

■ Exercise

Figure 7.1 shows René Magritte's 1929 painting of a pipe, which has the caption 'This is not a pipe.' Why is it not? What does this picture tell us about the nature of representation, and how might your answer to this have implications for the way in which we should think about geographical issues?

Figure 7.1 Ceci n'est pas une pipe (This is not a pipe), *The Treachery of Images* (1929), René Magritte

7.3 Place, space and knowledge

As we have begun to explore, representation employs systems of signs to create and convey meaning. To explain the relevance of these ideas for the study of geography, we might take the example of English landscape painting in the eighteenth and nineteenth century. A number of geographers (e.g. Cosgrove, 1985; Daniels, 1993; Nash, 1996) suggest that such paintings incorporate an **iconography**, a set of symbols which speak meaningfully to any observer who shares the cultural system of meaning that the painter is drawing upon. In this way, they suggest that it is possible to see all these representations of landscapes as '**texts**' which can be '**read**' by other people sharing broadly similar communicative systems of meaning. This representation of landscape has subsequently been interpreted as creating certain meanings of place in accordance with the interests of those cultural groups who had a vested interest in the maintenance of particular types of landscape. In short, the selective and stylized representation of landscape in British art has been seen as encouraging the preservation of certain types of landscape and not others.

Before considering this idea in more detail, we perhaps need to consider in more depth the relationship between representation, knowledge and social status. After all, if representation structures our understanding of the world, it is clearly linked to the construction of knowledge and the shaping of social relations between different cultural groups. An ability to represent things in certain ways (and not others) therefore becomes an ability to shape ideas about the type of relationships that exist between (for example) different classes, between men and women or between people and place. For instance, if a certain way of representing a place (or social relationship) via television, cinema and literature obscures other, alternative representations, it is perhaps unsurprising that this becomes the dominant way of thinking about that place (or relationship). Many geographers have therefore considered the relationships between representation and knowledge in terms of **power**.

While we will explore what is meant by power in more detail in Chapter 8, it is perhaps useful at this stage to introduce the ideas of the French philosopher and social historian Michel Foucault. In essence, Foucault (1981) suggested that knowledge is produced through **discourse**. Discourse is defined here as being similar to language, indicating a general way of talking about, thinking about or representing something. Two further points need making in conjunction with this: firstly, discourse is specific to particular times and places and, secondly, discourse is inextricably linked to **practice** (i.e. what people actually do). Through the idea of discourse then, we can begin to link meaning to practice. Both should perhaps be seen as *flows* – discourse as a flow of meaning and social practice as a flow of human actions. As Foucault explained, these flows serve to shape the workings of power in society; by controlling discourse, powerful groups in society are able to control, not just *what* people know, but *how* people know – the way people think about things. In contemporary Western society, for example, the scientific way of knowing about the world is a very powerful discourse which influences how we know and understand our place in the world. Scientific laws seemingly present an incontestable version of what exists and why it exists, with the 'boffin' becoming a figure whose authority is widely acknowledged (though sometimes parodied). But science is often linked to another powerful way of understanding and organizing the world – capitalist economics. This entails a set of business discourses which suggest that the world inevitably operates according to certain laws of the market. On the one hand, these discourses explain to us why the price of goods in our shops have gone up (again); on the other, they may explain away wealth inequalities, recessions, unemployment and poverty as 'natural' outcomes of the capitalist system.

This coincidence of scientific and capitalist discourses has two important results. Firstly, these powerful discourses of science and business become dominant ways of understanding the world. They become accepted as 'common sense', which makes them seem natural and difficult to question. As a result, the powerful groups and organizations (e.g. the wealthy, scientists, research institutes and big businesses) associated with the discourses of science and capitalism tend to remain powerful. The concept of **ideology** is significant here, having been defined by John Thompson (1990, 7) as 'meaning in the service of power'. What this means is that the powerful are able to manipulate and produce representations (in their broadest sense) so that particular meanings and understandings become or remain dominant. The current debate in the UK over the 'right to roam' serves as a brief illustration. For most landowners, it is seen as 'common sense' that control over parcels of land should lie with individual landowners, and that access to that land should be restricted. Ramblers' demands for a right to roam over the land are thus seen as far more than a desire to go for a walk: they can be represented by landowners as a challenge to the notion of private control and ownership which, ultimately, is central to capitalist economics. Despite recent moves towards legislating for a limited right to roam on uncultivated land, the powerful idea of private property is maintained. Alternatives like the nationalization of land or an unfettered right to roam are almost unthinkable in the face of such dominant discourses.

A second consequence of the ubiquity of these scientific and business discourses is that alternative ways of understanding the world are regarded as less important in comparison. If scientific ways of understanding society and the natural world are dominant, then it seems 'common sense' to define and deal with social or environmental issues through the application of scientific methods. For example, in responding to the Chernobyl nuclear accident of 1986, which affected some upland parts of Cumbria and Wales, scientists attempted to measure the effects of sheep grazing on radiation levels in vegetation by using standardized scientific methods (e.g. grazing a set number of sheep on a number of plots of equal size). Local sheep farmers, however, knew that microscale variations in relief, drainage and vegetation meant that the comparisons between the plots would be scientifically invalid, yet the scientists did not take the farmers' 'non-scientific' knowledge into account (Wynne, 1996). Scientific discourse, as a way of understanding the world, is often dismissive of other ways of understanding, rejecting them as subjective and emotional (rather than objective, rational and hard-headed). Adopting a 'Foucauldian' perspective on this (i.e. adopting Foucault's ideas), we can perhaps see that for every powerful and widely accepted discourse there are competing ways of looking at the world. Representation, therefore, can be involved in power struggles as it is caught up in the production of particular ways of looking at the world.

What we can conclude from this is that representations of people and places are tied to discourses which encompass particular ways of looking at the world (and not others). In turn, these discourses shape social practices. From this perspective, representation can be seen as a *process* shaping the organization of the world, implying that geographers always need to be aware of who is producing the representation (as well as who is consuming it). Another example might serve to emphasize this. Between 1975 and 1999 the Indonesian state annexed East Timor and sought to represent it as a legitimate part of Indonesia through government-owned media. Such representation was used to justify the oppression of any opposition to the Indonesian annexation. Accordingly, alternative representations of East Timor as a colonized, invaded territory were suppressed through violence, and it was only when images of violence in East Timor leaked out via foreign media that the international community mobilized to end this annexation. You can probably think of many other examples, at many different scales where similar processes of selective representation have been used to justify violence, terrorism and oppression. What is, of course, crucial here is that dominant representations cannot be taken at face value – they have been produced by particular groups for particular reasons. We should therefore remain mindful that for every dominant representation we encounter there potentially exists a host of **subordinate** representations.

■ Reading

For a good discussion of representation and power, see Stuart Hall (1997) 'The work of representation', in Hall 1997 (ed.) *Representation: Cultural Representations and Signifying Practices*. There is also a large literature that deals with the role of the media in producing and managing news that is informative for showing how representation influences our knowledge of what exists in the world. For examples that are inherently geographical, look at Burgess (1985) on the selective coverage of inner-city riots or Alderman (1997) on the global coverage of the O. J. Simpson case.

Let us now return to our consideration of English landscape art in the light of these ideas. On one level we might read landscape art as showing us the 'truth' of what type of scenery existed in the eighteenth or nineteenth century. Against this, some geographers have argued for a more critical reading that rejects this mimetic notion in favour of a deeper reading. Following their lead, we can perhaps look beyond the surface image to examine the ways in which the countryside as an *idea* was produced through such representations, and the way that this worked in favour of certain groups. Figure 7.2 shows Thomas Gainsborough's painting *Mr and Mrs Andrews*, painted around 1750. As you look at this, you might think about how the countryside is represented in this image. Maybe your initial impression is that this is a picturesque landscape, a peaceful, quiet place or a place of leisure and relaxation. Yet this is just one interpretation of the eighteenth-century countryside, and if we start to look beneath the surface of the picture, to 'read' it as a text, we can start to pull out some deeper ideas about the process of representation which contributes to the formation of particular ideas of the countryside. In doing this, what is *not* included in the representation may be as significant, or even more significant, than what is included. For example, the painting shows an agricultural landscape; there is clear evidence of hard agricultural labour. Yet the source of that labour, agricultural workers, is missing. In a sense, the non-landowning labourer is excluded from this private landscape by the painter. The effect is to uphold the ideal of private enjoyment – the labourer is allowed to do his or her work, but once that is done the landscape is not his or hers to enjoy. In fact, there were often severe penalties for trespass and suspected poaching. Further, the actual backbreaking agricultural labour which produced the landscape is not shown. What is shown is a landscape of leisure for the privileged few (including the painter himself, the landowners and the intended audience for the paintings).

There are other selective aspects of this form of landscape representation. For example, *Mr and Mrs Andrews* certainly celebrates ownership of property – land in this case. It also illustrates unequal gender relations – Mr Andrews is standing, in a

Figure 7.2 *Mr and Mrs Andrews*, Thomas Gainsborough (1748)

Photo: © National Gallery, London

position of authority, and is prepared for hunting, while the seated Mrs Andrews is in a far more passive position. At the time of painting, Mrs Andrews would effectively have been regarded as the property of Mr Andrews (so land is not the only thing being owned here), and her function would be to reproduce, to continue Mr Andrews' family line, and to provide an eventual heir to the landed property. Indeed, it is not the couple which owns the land, it is only Mr Andrews who is the proprietor. The painting, then, as a representation of the rural landscape, does far more than simply reflect the scene in front of the artist. Instead, it illustrates particular sets of social and power relationships, and through this illustration acts to reproduce (i.e. continue) those relationships (Rose, 1993). The *idea* of the countryside is itself created through these representations. The countryside here is owned, private and masculine – a space to be enjoyed by elite social groups. These representations are only *partial*, from the perspective of elite, wealthy, powerful groups in eighteenth- and nineteenth-century English society. From the perspective of the rural labouring poor, the countryside might have looked very different, a site of hard work and oppression.

■ Exercise

Towards the end of the nineteenth century photography became an increasingly popular way of creating images of the countryside and 'country life' (Figure 7.3). Think about the differences between *paintings* of the countryside and *photographs* of the countryside, as representations. What are the key similarities and differences between them as modes of representation? Are photographs necessarily more 'real' than paintings? What can we learn about the countryside, rural people and rural cultures from paintings and photographs?

Figure 7.3 Rural idylls? Early photography, from *Grundy's English Views* (1857)

Photos: Hulton Getty

Significant changes in English society since the rapid industrialization of the nineteenth century have undoubtedly contributed to important new forms of representation of the countryside. In particular, Britain has become a predominantly urban society, and the manifestations of this include the way in which large numbers of people are distanced from the countryside in their work and domestic life. Fewer people now work in agriculture, and most people spend most of their time in urban and suburban environments. As discussed in Chapter 5, their experience of urban places can be one of rapid, even frightening, change. As a result, there has arisen a very powerful set of social myths concerning the stability of countryside in relation to the dynamism of the city (see Chapter 6). The idea of the **rural idyll** is an important consequence of the way the contemporary countryside is represented (Mingay, 1989). Some key elements of this myth include:

- an impression of timelessness;
- an emphasis on traditional 'family' and community values;
- harmonious relations between 'nature' and 'culture';
- an absence of social problems;
- the fostering of good physical, spiritual and moral health.

In sum, the countryside is imagined to be a more pleasant place to be and live than the city. This image of the 'chocolate-box countryside' (a term deriving from the frequent use of reproductions of paintings like Constable's *Haywain* as wrappers for confectionery) has become very important in a predominantly urban society. What we again need to stress here is that the idyll is a selective representation of the countryside, imagined mainly from an urban perspective. This is not to say that it is *only* a representation and not important, as the idea of the rural idyll is highly significant in influencing the ways in which people enter into relationships with the countryside (for example by moving there to live, or by taking part in leisure activities from rambling to fox-hunting, from off-road driving to paintball war games).

Another important facet of the rural idyll is that it has been strongly associated with particular social groups in the UK. It has been suggested by many researchers that this particular representation of the countryside is linked to a group of people termed the 'service class' (Savage and Warde, 1993). This class consists of people in professional and managerial jobs (and their families), who are wealthy enough to be able to 'buy into' the rural idyll, most significantly through the purchase of a rural property (the process of *counterurbanization*), but also in other ways, such as the purchase of items which symbolize supposedly rural lifestyles (e.g. the stove and waxed jackets in Figure 7.4). The lifestyle of the service class encourages a process of **commodification**, whereby things which can be associated with the countryside, however tenuously, are turned into products to be bought and sold.

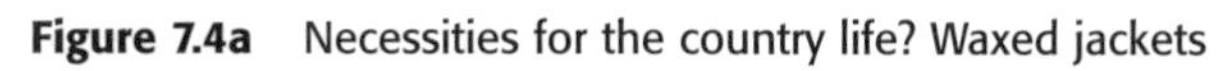

Figure 7.4a Necessities for the country life? Waxed jackets

Photo: Reproduced courtesy of J. Barbour & Sons Ltd

Figure 7.4b Necessities for the country life? Solid-fuel stoves

Photo: Stanley Cookers (GB) Ltd

Part of the process of representing the countryside, then, is the production of objects and images designed to be attractive to potential purchasers. Figure 7.5 shows a scene from Hatton Country World in South Warwickshire. Hatton consists of a range of old farm buildings converted into craft shops, factory shops offering discount clothing, and restaurants. The site also contains such features as a farm park and children's play area. Associations are continually drawn here between the countryside and what is on offer for sale: the crafts are labelled as 'rural' crafts; a restaurant is called 'The Greedy Pig'. At the same time, there is an assumption that what there is at Hatton is representative of the 'real' countryside: so that the farm park (with its pets' corner and guinea pig village) supposedly reflects agricultural production in the wider countryside. The countryside is represented at Hatton as a series of **spectacles** to look at, and hopefully (as far as the proprietors are concerned) to be purchased. More generally, many TV dramas, such as 'Last of the Summer Wine' and 'Heartbeat', draw on idyllic representations of the English countryside and country life to attract large viewing audiences (see Mordue, 1999). Ultimately, entire areas of the countryside may themselves be commodified, packaged and sold as 'Last of the Summer Wine' country or 'James Herriot' country in order to attract tourists.

However, as with the eighteenth- and nineteenth-century representations we examined earlier, we can look beneath the surface of contemporary idyllic representations of the countryside. Several issues emerge if this is done. Firstly, the representation of the English countryside as a rural idyll is one that excludes certain groups (who may find themselves feeling 'out of place' in the rural). Kinsman (1995), for example, has suggested that black and Asian people (for instance) may experience feelings of exclusion from an English countryside that is represented as a 'white' space. Secondly, the representation of the countryside as an unchanging, peaceful, harmonious idyll acts to cover up the ways in which the countryside can be a contested space. Ongoing debates in the UK about issues such as fox-hunting and the 'right to roam' illustrate that what is considered right and appropriate in the countryside is certainly not timeless, but that there are a variety of opinions about, for example, what is acceptable behaviour in the countryside (see Cox, 1993 and Cox and Winter, 1997, on the hunting issue). Thirdly, the idyllic representation of the countryside hides alternative experiences of rural life. For example, the chocolate-box imagery obscures the existence of poverty and deprivation in the countryside, while the farm park at Hatton Country World does not tell the stories of intensive livestock production or genetic engineering of crop plants. Dominant representations of space, such as the rural idyll, should not lead us to ignore alternative representations which may, for example, suggest that the countryside is a place of exclusion, poverty or simply boredom.

Figure 7.5 Rural commodification: Hatton Country World

■ Exercise

Figure 7.6 shows an advert for livestock feed designed for 'hobby' farmers, while Figure 7.7 is a newspaper photograph of the 'Countryside March', held in London in March 1998. Compare the way in which the advert represents the countryside with the ways in which the people in the photograph are representing the countryside. What are the reasons for representing the countryside in these particular ways? The newspaper photograph is itself a representation of an event; think about the role newspapers play in representing events such as the 'Countryside March', and consider how the marchers might have planned their own representation of the countryside in response to anticipated newspaper coverage. You might also compare these representations with the promotion of Hatton Country World.

While our understanding of many rural issues is not solely based on representation, you should be able to think about how newspapers could present a positive representation of fox-hunting (that it is a natural part of rural life) as easily as a negative one (that it is a cruel pastime). The media, therefore, plays a crucial role in the production of place myths. Equally, intensive farming can be represented as bad (it doesn't allow animals to express their 'natural' behaviour) or good (it provides animals with secure environments and humans with cheap food). The countryside itself can similarly be represented in different ways, depending on who is doing the representing. For this reason, suggesting that there is a rural 'reality' is problematic. For example, the countryside can be represented as a place of leisure by ramblers, and simultaneously as a place of food-production by farmers. The farmers may themselves be represented as selfish and greedy by the ramblers (the 'get off my land' stereotype springs to mind), while they represent themselves as simply trying to make an honest living from the land. Representation, then, is a key part of the way in which people engage with and debate moral and ethical issues concerning what is happening and what should happen in particular places.

■ Reading

Interesting books on landscape painting and the representation of the English countryside include John Barrell's (1980) *The Dark Side of the Landscape*, Anne Bermingham's (1986) *Landscape and Ideology* and Stephen Daniels' (1993) *Fields of Vision*. Broader discussions of rural representations can be found in Michael Bunce's (1994) *The Countryside Ideal* and in the second chapter of John Short's (1991) *Imagined Country*. David Matless (1998) provides a focused account of the connections between landscape, rurality and modernity in the years 1918–51 in *Landscape and Englishness* (Chapter 1 is particularly recommended).

Figure 7.6 'Especially for the Small Holder': a farming idyll

Source: Allen & Page Specialist Feeds

Figure 7.7 The countryside comes to town: Countryside March, March 1998, London

Photo: © Peter Arkell/Impact Photos

7.4 'It's grim up North . . .': representing regions

Inevitably, place myths need to be understood in relation to the representations that exist on a variety of different spatial scales. Our interpretation of the English countryside (above) can only really be understood in relation to wider representations of English life; while representations of specific villages intersect with representations of Englishness, a Dorset village may be represented and imagined as very different from a Yorkshire village, though both are established as quintessentially English in their own way. In this sense, representations of nationhood are always internally differentiated, invoking ideas of specific **regional** cultures and ways of life. Our second major example thus focuses on representations of 'North' and 'South' in England. As we will see, representations of these two imaginary regions are tied to a long history of uneven development. Although it is possible to identify other regional divides, the persistence of the 'North–South divide' suggests important social, economic and cultural rifts, perpetuating distinctive stereotypes of people from both regions. In this section, we look briefly at the nature of this divide, examining the role of representation in its reproduction, and, linking back to Chapter 6, we want to suggest that the English North and South exist as much as *imagined* places as they do as 'real' places.

Rob Shields (1991) has offered a detailed analysis of the ways in which the English North has been imagined over the last 150 years. In his assessment, it is important to realize that the Industrial Revolution had greater immediate impacts on the North (including the English Midlands) than on the South, with the development of manufacturing and heavy industry and the establishment of company-dominated towns. However, as such industry declined in importance (particularly from the 1970s onwards) and new jobs were created in the service sector (especially financial and creative services), it is suggested that an economic and social divide between North and South emerged. One of the key manifestations of this is higher unemployment and greater social deprivation in the North than in the South. However, Shields argues that this division between North and South in England goes further and deeper than this (admittedly important) social and economic divide. He suggests that while the origins of the North–South divide have a material basis (i.e. that their roots lie in significant differences in the conditions of wealth-production) the idea of a divide has been fostered through the ways in which the Southern-dominated media have represented the North in novels, films and television.

Here, it is necessary to trace the origins of this process back to the Industrial Revolution, beginning in the mid-eighteenth century. This fundamentally (if temporarily) shifted the location of economic power in England. Previously, power had been located in the South, based on an aristocratic landowning system (where land = wealth = power), agriculture, and the centralization of administrative and royal power in London. With the Industrial Revolution, a new economic system developed based on resources (e.g. coal and water power) located in the North and

Midlands. The North, then, increased in importance as a centre of wealth. This wealth, however, was different to the 'old' wealth of the South, and was associated with important new places and people. Firstly, it was wealth generated from manufacturing processes, such as weaving, rather than being inherited and tied to landownership. The new wealthy people were also different, being industrialists rather than landed gentry and aristocrats. A situation had developed where the power of an old elite was threatened by newly wealthy and powerful industrialists operating in different, new types of place: factories. The factory system of mass production and a highly regulated workforce produced a new social order (Hetherington, 1997), something completely different to previous small-scale, cottage-production systems. Accordingly (and somewhat hypocritically), the old elite despised the emerging new industrial elites for the vulgarity of making money through trade and industry. This in itself is a key part of the development of the North–South divide, with a snobbish Southern elite representing Northern industrialists as vulgar venture capitalists who favoured wealth-accumulation and hard work over the values of the landed and leisured aristocracy.

It is apparent here that the imaging of the North stemmed from the South as a way of confirming the superiority of Southern values (Wiener, 1981). This type of analysis can be carried forward into a broader consideration of stereotypical elements of North and South. Evidently, the Industrial Revolution came to be associated with particular urban forms and social and economic conditions, set within a geographical area with a particular physical character. These aspects combined to form stereotyped visions of Northern landscapes, Northern cities and Northern people. In opposition to the pastoral, gentle landscapes of the South, the Northern landscape has been characterized as grim, hellish, bleak, cold and rugged. This was a new industrial landscape consisting of factories and their associated chimneys and pollution, and new forms of housing – in particular, large areas of terraces. This urban landscape was imagined by Southerners as ugly and functional, as wholly practical rather than aesthetically pleasing (see Figure 7.8). There is a contrast here to the supposed rurality of the South, and the development of Southern cities (and the slum areas of London, for example) is ignored in the representation of these mythical landscapes. Finally, apart from the new industrial elite, most of the people in this new urban landscape were certainly not wealthy or powerful; most people in the new industries were workers, including children. Most worked long hours for low pay, while the industrialists got rich through the exploitation and control of the workers. There was, then, the emergence of a new 'working class', which has continued to play an important part in the stereotypical representation of the North. Working-class Northern men, for example, are continually (if humorously) associated with pigeons, whippets and flat caps. The mythical Northern working class is also strongly associated with ideas of close-knit community.

This imagined North was reflected in, and constructed by, contemporary representations such as the following description of the fictional Coketown in *Hard Times* by Charles Dickens in 1854 (quoted in Shields, 1991, 210–11):

Figure 7.8 A working townscape: Burslem, Staffordshire

Photo: Aerofilms Limited

> It was a town of red brick, or of brick that would have been red if the smoke and ashes had allowed it; but as matters stood it was a town of unnatural black and red like the painted face of a savage. It was a town of machinery and tall chimneys, out of which interminable serpents of smoke trailed themselves for ever and ever . . . . It had a black canal in it, and a river that ran purple with ill-smelling dye, and vast piles of buildings full of windows where there was a rattling and trembling all day long . . . . You saw nothing in Coketown but what was severely workful . . . . The jail might have been the infirmary, the infirmary might have been the jail, the town-hall might have been either, or both, or anything else.

The basis for this is Preston in Lancashire, in the mid-nineteenth century, but the description clearly draws upon, and reproduces, a stereotypical vision of an austere, grim and ugly urban industrial landscape. Dickens himself was an author who wrote most of his books while living in Kent or London, and from his Southern perspective the North was 'the pole against which the civilisation of the South was compared' (Shields, 1991, 230). Table 7.1 illustrates some of the ways in which this opposition of North and South developed through images of landscape, lifestyle and industry.

Alongside these myths has been the concurrent development of a so-called 'cult of Northern-ness' (i.e. stereotypes of Northern-ness and Southern-ness written from a *Northern* perspective). Here, the North represents itself as superior to (and opposed to) the South; the rugged, honest, hard-working and kind-hearted Northerner is contrasted with the 'soft', lazy and snobbish Southerner. Clearly, this is a stereotypical representation – based on partial representation – but it has affected social practice. For example, during the industrial revolution, the Northern industrialist would frequently aspire to a country estate, and so gain the trappings of inherited wealth, because of the idea that earned money was less culturally valuable than inherited money. In more recent times, stereotypes of the North and South may be seen as influential in shaping patterns of investment. Here is one leading UK businessman, talking about the different markets for his product in the UK:

> I would like you to picture an imaginary line from the Bristol Channel to the Wash. Above that line we have the beer-drinking, chip-eating, council house-dwelling masses, probably with lower car ownership. Below . . . are wine-drinking, courgette and mange tout-eating, semi-detached dwelling [southerners].
>
> Reported in the *Guardian*, 5 January 1998.

So, despite their origins in the past, these stereotypical representations of North and South in England are still significant in the present. Recent films like *Brassed-Off*, *The Full Monty*, *When Saturday Comes* and *Rita, Sue and Bob Too*, novels such as *A Kestrel for a Knave* (Barry Hines, 1968) and *A Kind of Loving* (Stan Barstow, 1960) and numerous television programmes (such as 'Coronation Street', 'Emmerdale Farm', 'Heartbeat', 'Byker Grove', 'Last of the Summer Wine', 'Band of Gold' or 'Brass') focus on stereotypical people, places and landscapes of the North.

Table 7.1 Images of North and South

'North'	'South'
marginal, peripheral	central, hub
working-class	economic and political elites
bleak landscape	tame landscape
manufacturing industry	service industry
rugged leisure pursuits	'high' culture (the arts)
wet and cold	milder climate
community orientation	individual orientation

Source: Adapted from Shields (1991)

Stereotypes of Northern people also have a place in advertising. Research has indicated that some Northern accents are more effective in persuading people to buy what is being advertised, suggesting that these accents signify trustworthiness, honesty and straightforwardness. These, in adverts for products like Hovis bread and Tetley tea bags, are associated with working-class and community-oriented values. Other adverts, such as those for Boddingtons beer (traditionally brewed in Manchester), are more playful with the stereotypes, using them ironically and humorously.

A very different representation of the North can be found in a new generation of themed museums, such as the Beamish Open Air Museum in County Durham and Kelham Industrial Island in Sheffield. These invite visitors to 'step back into the past', using multimedia displays, re-enactment and visitor interaction to create a participatory learning experience. These invariably contribute to and maintain some of the myths of the North by representing it as a place of heavy industry bound together by a close community spirit. However, what visitors see at such museums is largely **sanitized** and **selective**, in the sense that exact replications of the experience of Northern working-class life during the nineteenth or early twentieth centuries are impossible – the visitor knows that outside the museum the modern world continues, and strict health and safety laws prohibit practices which would have been common in the past. Some commentators contend that these 'living' museums act to **romanticize** Northern working-class life, suggesting that although the work was hard there was strong 'community spirit'. What is less often included in these representations is the existence of insecurity, low pay, illness, hunger, industrial accidents and poor housing. Admittedly these things might be mentioned, but they rarely become part of the experience of the museum in the same way that having a ride in a horse-drawn tram does. John Urry (1990) has thus argued that heritage sites, by offering costume drama and re-enactment, tend to trivialize the variety of social experiences connected with the past, making it safe and sterile. Similarly, Hewison (1987) refers to heritage representation as 'bogus history', carefully designed not to raise difficult questions and conforming to a uniform soft-focus image of Northern life. More widely, many heritage sites are accused of celebrating 'the great men of history' while overlooking the histories of women; likewise, children, ethnic minorities and the disabled are groups that usually only feature in these representations supporting the heroes at the centre (Shurmer-Smith and Hannam, 1994).

An important point to make here is that attempts to rework images of Northernness in museums and the like are often a central plank of attempts to attract new investment to Northern towns and cities. Such approaches are particularly associated with those previously affluent industrial cities where so-called 'rustbelt' industries, such as steel, vehicle-manufacturing, textiles and chemicals have declined in the face of foreign competition and new corporate investment strategies. Under such traumatic circumstances, cities like Manchester, Glasgow, Liverpool and Newcastle have been left with little choice but to compete for increasingly footloose capital (see

Chapter 2) by emphasizing their distinctive social, cultural and physical characteristics, inevitably working *against* inherited notions that they are grim, cold and unattractive places to live and work. Consequently, these cities have sought to reinvent themselves as successful post-industrial cities boasting a distinctive cultural heritage, a skilled population and good quality of life. This has involved sometimes elaborate promotion efforts with a bewildering array of leaflets, posters, videos, brochures and Internet sites used to extol the virtues of particular cities (see Figure 7.9). Of course, the selling of cities is not a new phenomenon, and, in the British context, can be traced back to the civic boosterism practised by the Victorian city fathers. Yet contemporary forms of place promotion must be seen as qualitatively different from these early approaches; rather than simply advertising the city, current approaches to place-marketing typically try to 're-image' or reinvent the city, weaving place myths which are designed to make the city seem attractive as a site for external investment. Short (1996) suggests the key myths here are that the city

Figure 7.9 Promoting place: Bradford, West Yorkshire

is multicultural (and accepting of difference), environmentally friendly, rich in cultural attractions and supportive of new investors – or, to put it another way, that it is simultaneously a fun city, a green city, a pluralist city and a business city (see also Short and Kim, 1998).

In this sense, **place-marketing** is a key strategy pursued by urban governors in an attempt to make the city more attractive to potential investors. This conscious manipulation and promotion of city imagery has been the object of considerable attention by researchers (e.g. Kearns and Philo, 1994), with much cynicism evident as to the arbitrary and selective way cities are promoted. For instance, Boyle and Hughes (1995) contend that Glasgow's reinvention as European City of Culture through the 'Glasgow's Miles Better' campaign and promotion of local architects and artists like Charles Rennie Mackintosh was one that alienated much of the city's population. Specifically, it was seen by many as a denial of the city's working-class heritage because of its failure to celebrate (for example) the contribution of the shipbuilding industry to the local economy, excluding industrial representations of the city in favour of a glossy, post-industrial vision. Subsequent counter-representations sought to redress this balance, aiming to shatter the newly manufactured image of the city in favour of one that better represented the lifestyles and experiences of Glasgow's working-class populace. Equally, in other cities it is possible to discern that the hyperbole of place-marketeers often ignores the contribution of specific social groups and neighbourhoods to the city in favour of a 'soft-focus' imagery designed to appeal to white, middle-class, heterosexual, able-bodied consumers and investors. Again, what this demonstrates is that representations are rarely innocent, being enmeshed in a complex **cultural politics** by promoting certain senses of place in favour of others. Appreciating whose images and ideas are being enshrined in representations of place – and whose are excluded – is therefore crucial in interpreting the meaning of place imagery.

A final point that we might make here is that attempts to reverse images of decline are rarely limited to the launch of a new advertising campaign, and frequently go hand in hand with the physical creation of a new urban landscape. The construction and promotion of 'spectacular' new urban landscapes, typically in derelict or waterfront areas, has been an almost universal response to de-industrialization in British and American cities, following the success of the rejuvenation of Baltimore's waterfront in the 1960s. Such redevelopments frequently centre on a 'flagship' project, a term commonly applied to a pioneering, large-scale urban renewal project such as a cultural centre, conference suite or heritage park designed to play an influential and catalytic role in urban regeneration. Inevitably, this redevelopment and 'repackaging' of urban districts into areas of (and for) consumption is heavily promoted by urban governors, in effect becoming a representation of the city in its own right (Hubbard, 1995). The design of these new packaged landscapes has therefore been proposed to be a crucial instrument in the orchestrated production of a new post-industrial city image, which, if successful, can create a sense of civic pride and loyalty to place (Harvey, 1989). Here then, we might start to think about the ability of the landscape to act as a representation of itself, promoting particular images of the city through its architectural appearance.

■ Reading

The key reading here is Chapter 5 of Rob Shields' (1991) book, *Places on the Margin*, on 'The North–South divide in England'. On the way that Northern cities are trying to reinvent themselves in the face of these pernicious place myths, see Watson (1991). More specifically, there is now a lively literature on the way that museums and heritage parks represent particular versions or 'stories' of the past, and we recommend the work of Hewison (1987) as an entertaining entry point into this literature.

7.5 Geography as representation: maps and map-making

In the final section of this chapter, we want to examine the way in which geography, as an academic subject, is itself very much in the business of representation. In theorizing and writing about what happens in particular places, geographers construct academic representations (in books like this, for example). Geographers are therefore engaged in the creation of place through the process of representation, in the same way that other groups create place via representation (indeed, human geographers frequently draw upon non-academic representations in their own representations). This implies that geographical representation of place is itself partial and selective, with a continuous dialogue between geographers about *how* particular places can best be represented (see especially Barnes and Duncan, 1992). In response to this dialogue, distinctive ways of representing the world have been adopted by geographers over the last hundred years or so. For instance, in Chapter 1 we touched on geography's reinvention as a spatial science. This was based on a belief that the world could be understood – and faithfully represented – using formal concepts and ideas derived from maths and science. Here, formulae, graphs and equations stood for (represented) the relationship between people and place. As we have already indicated, this is a form of representation with which many geographers have subsequently felt very uncomfortable, detached as it is from the ways in which people look at, make sense of and feel about their surroundings. In more recent times, geographers have therefore begun to explore new ways of studying and representing the relationship that exists between people and place, recognizing that their descriptions of place always present a partial view of complex realities (a theme that we will return to in Chapter 10).

Throughout these discussions and debates about representation, geographers have nonetheless continued to be associated with one particular way of representing the world – mapping. For most people, maps are imbued with forms of authority and 'official' maps (like those of the British Ordnance Survey) are commonly taken to be true reflections of the landscape they cover (i.e. they are considered as somehow 'real'). But whether we consider the sketch-maps that people create on the basis of their mental images (see Chapter 3) or the OS maps based on hours of fastidious surveying, all maps are partial, selective and, to some extent, distorted. As Roszak (1972, 408; quoted in Wood, 1993) puts it: 'we forfeit the whole value of a map if we forget that it is *not* the landscape itself, or anything like an exhaustive picture of it'. Instead, of course, maps are created by people as representations of landscape, and because of that we need to think of them as **socially constructed**. Maps are created by particular groups of people, for particular reasons, so that while most people are familiar with the idea of reading a map in order to find their way, we can also think about reading maps *critically*, examining the conditions in which they are produced. This means that we can read something about a society from the maps it produces and uses. Like other representations, maps can tell us about power relations, and about a group of people's way of understanding the world. By looking at a map and exploring what lies behind it, we can get some idea of how members of that society view the world and their place in it.

For instance, a common feature of many maps of the world has been the placing at the centre of what is important to the people producing them. Many ancient maps, such as those associated with the Mayan society of Mexico or early Christian maps of the world such as the *mappae mundi* (e.g. Figure 7.10) tend to compress space and time, and to aim to tell stories about creation, mythical gods or religious beliefs. The Ebstorf map, for example, places the Christian Holy Land at the centre of the world. The effect of this is to structure the outlook of those who see the map; representing the world in particular ways acts to influence people's perceptions so that religious beliefs, for example, are reproduced. By placing Christian authority at the centre, this type of map emphasized that people's place in the world was god-given and that their behaviour was subject to divine authority (Sennett, 1994).

At the same time as centralizing some things, maps always have edges – there are necessarily things which are on the margins (or even off the map altogether). In drawing maps, cartographers have had to develop techniques for dealing with things which are unknown or marginalized, things which are 'other' (see Chapters 4 and 6). This can be seen as a way of thinking about group identities – defining what is 'self' and 'other'. On ancient maps, for example, unknown people were frequently represented as grotesque monsters or freaks, or simply barbarians, living in strange and frightening places (Sibley, 1995). Such maps were dealing with people and places which were outside the experience of the people producing them.

This notion of (mapped) centrality and peripherality is also significant in thinking about international politics and power relations. As a result of European global exploration and imperialism, most contemporary maps have developed from a European perspective and place Europe at the centre. Although we know that the world is approximately spherical, a Eurocentric world map structures our outlook on the world in important ways. Firstly, if Europe is central, then non-European

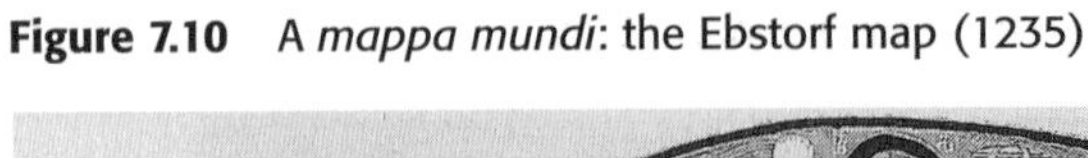

Figure 7.10 A *mappa mundi*: the Ebstorf map (1235)

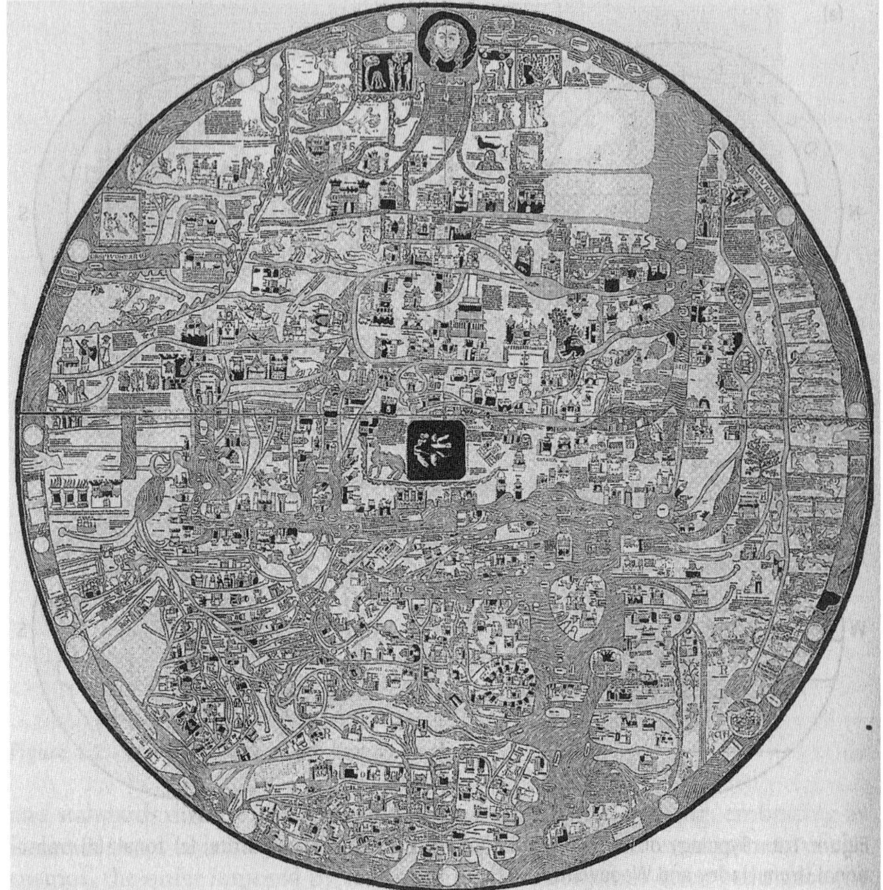

Source: Dorling and Fairbairn (1997)

places are at the edge, which makes them seem less important. Secondly, the map projection used in this type of representation makes the African continent, for example, look relatively smaller than it is, while Europe looks relatively bigger. European perceptions of Europe as important are both reflected in and confirmed by this type of map. This has increasingly been recognized in other parts of the world, and alternative maps have been produced which attempt to restructure views of the world, and make people think about some of the assumptions contained within Eurocentric maps (e.g. Figure 7.11), while early Christian maps often defined places as Christian or not Christian, images of places which are 'not us' are continually reproduced through this sort of mapping practice, so that, for example, during the 'Cold War' Western countries could represent the Soviet Union as a 'red menace' by shading the countries of the Soviet Bloc in red.

Figure 7.11 Redrawing the world: the polar azimuthal projection

Source: Dorling and Fairbairn (1997)

As well as defining **geopolitical** relationships between the society or nation producing a map and those outside of that society, maps are also implicated in the unequal social relations that occur within societies. Inevitably, all maps say things about the social order of a society, and about its historical development, and while it may be traditional for a government or tourist map to incorporate certain features, it neglects to include things that are no doubt significant to many.

■ Exercise

Figure 7.12 is a fragment of an OS map of an area on the English/Welsh border. From an examination of this map, what can you say about the UK's society and history? What things are missing from this map? What can the map tell you about the operation of power relations in the landscape it represents? From what perspectives is this map drawn?

Your reaction to this exercise may be influenced by how familiar you are with this type of map. If you have not seen or used this type of mapping before, you will perhaps be unsure as to what the various symbols stand for. On the other hand, you may routinely refer to similar maps in your working or leisure time, and find it difficult to think about how they reflect certain ways of looking at the world. For us though, important ideas emerging from this exercise include:

Figure 7.12 A fragment of OS map

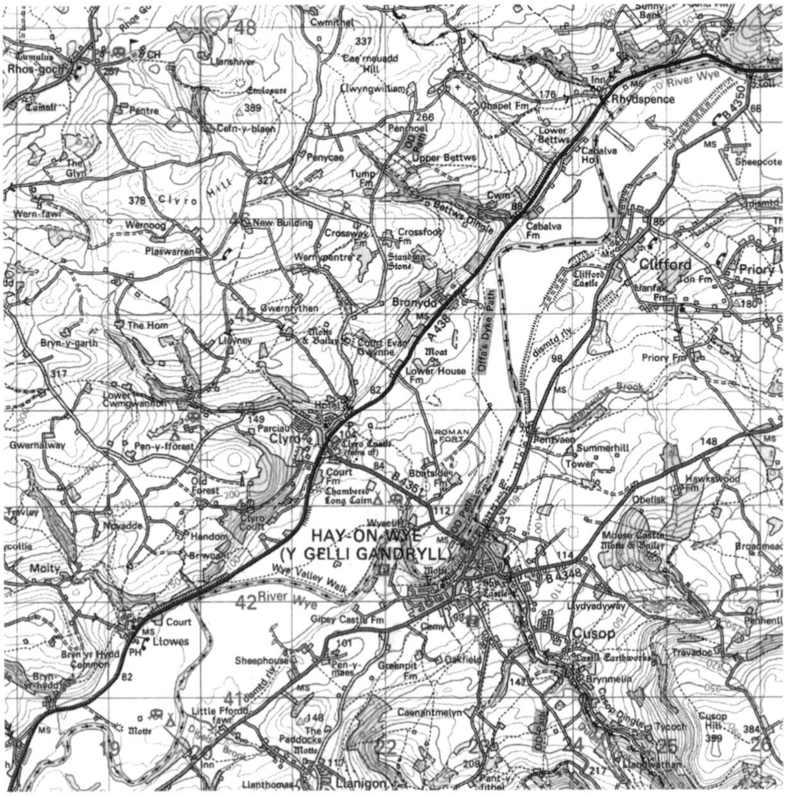

Source: © Crown copyright

- Particular things considered important in British society – such as property, transport routes and social/tourist infrastructure – are regarded as worthy of inclusion, and this is reflected in the prominence given to these features on the map.
- What is as important here, is what is not shown. By way of illustration, the map shows many relict features such as standing stores and defensive earthworks, but not old factories. This demonstrates several things. First, it provides a particular account of the historical development of the UK, suggesting that certain objects and events are more important than others. Second, it reminds us that maps are not just reflections of reality, but shape that reality, encouraging people to visit some locations but not others.
- The way in which a society thinks about itself is reflected in the composition of the map. Mapping carried out by the Ordnance Survey, for example, is highly *scientific*, its accuracy and precision reflected in the grid that overlays the map. As mentioned previously, science is something which itself plays a key role in our understanding of the world in general, and the OS map is a reflection of the significance of science in our lives. This can be seen as part of a power relation where OS maps are regarded as 'official' maps and others as inferior. Similarly, the vast 'non-scientific' range of human experiences of and feelings about places are not incorporated into OS maps.
- The existence of power relations is illustrated on maps by the existence of hierarchies and by the prominence given to particular features. For example, cities are given larger letters than towns, which in turn are given larger letters than villages. This seems like common sense, but at the same time it reproduces ways of thinking about settlement size which assume that cities are more important than villages. Similarly, the inclusion of certain administrative or political boundaries symbolizes the authority of the state. Such boundaries, like emotions and experiences, are clearly not 'naturally' present on the landsurface (although some may have been marked by people, with fences or stones, etc.), but are central to the government of society and the operation of power, and hence are displayed on maps. You might also like to consider the way in which English place names are prioritized over their Welsh alternatives.

In a variety of ways, then, maps say important things about the society which produces them. This idea can be further developed by examining the ways in which some maps come to be seen as better, or more valid, than others, and the ways in which groups of people use maps to control and define places. Maps, as representations, are not simply passive pictures but are entwined in flows of knowledge and power. As Dorling and Fairbairn (1997, 71) contend, 'a picture is a map when it is drawn by somebody with the authority to draw maps'. This suggests that powerful groups in society are able to control what counts as a map, and who is able (or authorized) to draw a valid or acceptable map. By doing so, alternative ways of viewing and mapping the world by may be subjugated and regarded as invalid. For example, the sketch-maps which adults routinely produce to help people find certain things may be disregarded as childish, unscientific and subjective when contrasted with the 'proper' maps of the Ordnance Survey.

This type of argument can be replicated for maps produced in different contexts and at different scales. To return to an earlier example, European imperialism relied on the use of maps to classify and order the colonized territories. In this context, mapping a territory (frequently imposing new boundaries and new names) was a way of taking possession of that territory, especially when the names and boundaries of the pre-existing society were taken off the map as if they never existed. In colonizing North America and Africa, for instance, Europeans used maps to justify colonization. These maps showed empty spaces and potential resources for exploitation but seldom showed things which were important to pre-existing societies (such as religious or mythical fields of care). For the Europeans, it was as if there was nothing there, so that they could take possession and claim land before it was even occupied. By doing this, European colonialists were claiming superiority by suggesting that other cultures were insignificant. Mapping, then, is a very powerful way of taking control of territory; by drawing boundaries on a map, for example, something is brought into existence, and an impression is given of fixity and permanence. This was backed up by implicitly saying that European ways of making maps were superior; these were the technical and scientific maps which Western society considers to be 'proper' maps. In some cases, where colonized people drew maps in a pictorial style, they were dismissed as 'child-like' (confirming the validity of Dorling and Fairbairn's comment).

The key point to stress here is that geographers should never take maps for granted. As powerful (and seemingly objective) representations of place, they need to be critically 'read' in the same ways as writing and painting to expose the basis on which they were authorized. To finish this section, we want to consider briefly the ways in which 'new' types of space are being represented by exploring the mapping of 'cyberspace'. Evidently, cyberspace and the Internet are often conceptualized in geographical terms (the word cyberspace itself contains an important geographical referent). The word 'space' is used here as a metaphor: that is, it is used imaginatively but not literally to think about the ways in which the Internet connects people and places. This has several implications. For instance, it demonstrates that humans want (or perhaps need) to think spatially or geographically. Thinking about the Internet spatially allows people to imagine it as having a 'geography' that they are able to navigate. However, as a conceptual space, cyberspace is free from many of the constraints of 'real' space – it is an imagined space in which extraordinary things can happen, it is unconstrained by many of the barriers which exist in 'real' space (although there may be other barriers in these new spaces, such as getting access to a computer terminal). Cyberspace is also unlike 'real' space (but perhaps similar to 'imagined' spaces) in that it is continually being constructed, changed and expanded as it is used. As such, attempts to map cyberspace using conventions of two-dimensional representations appear doomed to failure.

Representing the Internet as a series of spaces allows the use of many other imaginative spatial or geographical metaphors. Despite the fixed position of the computer terminal, we imagine *moving* and *surfing* through cyberspace, bringing new communities and new frontiers into existence. Being in cyberspace allows the construction of new identities in an imagined space of liberation and freedom of communication between people. People have the freedom to be who they want to be to a large extent, because they cannot be seen. It becomes possible to play different parts, to pretend to be something you are not or to explore parts of yourself which you normally cannot (though there are some caveats to this, as we will see in Chapter 9). At the same time, and despite the global nature and image of the Internet, there is a paradoxical need to stress the local, to place ourselves on this new map (people are located on their *home*pages).

However, while it is tempting to celebrate the exciting possibilities of cyberspace, we also need to note that not everyone has access to this **representational space**. Being in cyberspace requires specialist and expensive equipment, as well as specialized knowledge. Although computers, e-mail and the Internet are fairly common for some groups of people (especially in Western nations), for other groups (and in other parts of the world) these new spaces are inaccessible. At the same time, the commercial possibilities of the Internet are increasingly recognized – and commodified – by people wanting to advertise or sell in cyberspace. This means that maps of cyberspace may be used to help commercial organizations make sense of who is using the Internet (and accordingly influence their behaviour). Specifically, by recording some websites, servers or discussion groups (and not others), maps of cyberspace would have the ability to guide surfers to certain sites (and not others). This would mean that maps of cyberspace would potentially marginalize new groups seeking to gain a footing on the Internet; not to be on the map would mean that the group would be effectively invisible in the informational economy. In the same way that the companies responsible for creating Internet search engines (e.g. Lycos, Yahoo or Netscape) are becoming powerful for providing access to websites, a company that could come up with a seemingly definitive atlas of cyberspace would no doubt be hugely influential (as well as commercially successful).

At the moment, such an atlas is some way off (perhaps not surprisingly given the ephemeral nature of most sites). Yet these conceptual discussions about the way that we might represent the imagined spaces of the Internet are not entirely hypothetical, with many organizations currently seeking to create databases and directories of the Internet. Even though we think that it is important that geographers try to contribute to these discussions in their attempt to make sense of this new technology, we feel it is important that they remain as wary of using conventional cartographic concepts to represent this as they might when they map 'real' spaces. Ultimately, maps of both 'real' space and cyberspace are always selective, partial and distorted, serving the interests of some groups and not others. Geographers cannot distance themselves from these power relations – like everyone else, they are caught up in flows of discourse and practice when they create their representations of the world.

It is the complexity of these power relations that we want to expand on in the next chapter.

■ Reading

There is a wide range of interesting material examining mapping as a form of representation. Some key suggestions include Harley's chapter 'Maps, knowledge and power', in Cosgrove and Daniels (eds) (1988), *The Iconography of Landscape*; Doreen Massey's (1995b) chapter 'Imagining the world' in Allen and Massey (eds) *Geographical Worlds*, and Wood's (1993) fascinating book *The Power of Maps*. In addition, Dorling and Fairbairn's (1997) book, *Mapping: Ways of Representing the World*, is useful (especially Chapters 4, 5 and 8); they also critically evaluate Geographical Information Systems (GIS), suggesting that the specialized skills and knowledge required to operate such computer-based database and mapping systems can marginalize some groups, while the requirements of GIS might result in 'difficult' non-English place names being Anglicized (Chapters 4 and 7). Crouch and Matless (1996) examine alternative, 'lay' modes of mapping in relation to the creation of 'Parish Maps'. 'Geographies of cyberspace' are considered in the book of the same name by Kitchin (1997), while his paper in *Progress in Human Geography*, entitled 'Towards geographies of cyberspace' (1998) also offers a useful examination of how geographic ideas have been applied to the understanding of new technologies. More generally, Mitchell's (1996) book, *City of Bits*, offers an optimistic account of the potential of the Internet. Finally, Duncan and Ley (1993) provide a valuable and accessible discussion of how modes of geographical representation have changed over time.

■ Summary

This chapter has begun to explore the importance of representation in everyday life. Here, we have suggested that representation is central to the ways in which societies function, and can be linked to people's ways of making sense of the world. Specifically, we have argued that

- Representation relies on *languages* and *systems of meaning* which are used for communicating experience and knowledge of the world between members of cultural communities and groups.
- Representation requires constant interaction between people, who are involved in representing the world (and people and places in the world) to each other in a variety of ways for a wide range of reasons. Meaning is constantly created through processes of representation.

- It is possible to 'read' a whole range of representations (e.g. novels, paintings, TV programmes and maps) to explore what they can tell us about the society or individual which produces and consumes them.
- Thinking about the operation of power relations is central to understanding representation – representation is deeply involved with struggles to control people and places.

eight Place and power

This chapter covers:

8.1 Introduction

In previous chapters we have looked at the idea that people's relationships with the places they use and pass through can be usefully illuminated by focusing on the cultural and social values which are inscribed in those places. Ideas from human geographers about the importance of place in consolidating people's identities have been drawn on to explore the way in which places act as signification systems, in the process of constructing particular sets of meanings or myths. Implicit in these ideas is the assertion that people's relationships with their surroundings are shaped by the relationships that they have with other people, and, consequently, that their cultural identity can (to some extent) be 'read off' from the places that they occupy (and the representations they make of those places). As such, it is apparent that certain groups – whether defined in terms of their ethnicity, age, gender, sexuality or income – find themselves marginalized in wider **systems of space**, able to inscribe their values only in certain (and often marginal) locations. This implies that most 'everyday' landscapes can be read as the products of attempts by dominant cultural groups to inscribe their values into the geographic landscape; in most Western societies, as we have begun to demonstrate, this means that places often reflect the cultural values and interests of white, wealthy, male, bourgeois, heterosexual, ablebodied people. But many geographers suggest that this reading of place, whereby some cultural values are favoured over others, should not be taken for granted, as the 'natural' outcome of people's relationships with one another and the places they occupy. Instead, they have argued that it is the outcome of a complex **cultural politics** where certain values are incorporated into the mainstream and others are pushed to the margins as a result of the unequal power relations evident in Western societies (see especially Jackson, 1989).

In this chapter, then, we want to expand upon some of the ideas developed by geographers about the way in which places can encapsulate certain cultural values in the service of power. Though differing on specific points of emphasis, the ideas we explore here broadly argue that the relationships that people have with place cannot be considered without an analysis of power relations in society; or, to put it another way, that the design, organization and use of place reflects forms of social and cultural inequality. Here, the notion of power is taken to have a meaning somewhat different to the meaning that you may be accustomed to. Specifically, although power is frequently defined as meaning control by **authority**, most of the writers we draw upon in this chapter do not suggest that the relationship between people and place is controlled or determined simply by the authority of a few powerful individuals or cultural groups. Rather, they tend to suggest that the relationship between people and place itself is part of a broader and more diffuse exercise of power, and that we are all 'caught up' in complex power relationships. This is not necessarily an easy argument to grasp, and in this chapter it will also become apparent that power remains a disputed concept amongst geographers, seen variously as taking violent, coercive and persuasive forms (often all at once). Equally, it will become evident that power may be discerned in the relations that exist between people and places at all scales and levels of society, from individual household relations to the type of interdependencies that exist between the peoples of the more and less developed worlds. In this light, the starting point for our consideration of power and place – institutional settings of the nineteenth century – may initially seem an odd one. Yet, as we describe, this focus allows us to engage explicitly with the ideas of Michel Foucault, a figure whose spirit continues to loom large over any geographical discussion of power (and representation, see Chapter 7). In the remainder of the chapter, however, we move from the nineteenth century into different times and places in order to explore how ideas about power can help explain how a variety of everyday places are perceived, understood and used.

8.2 Power, discipline and the state

Traditionally, power has been written of as encompassing the ability of a dominant group to achieve certain ends (e.g. to maintain its dominance, or perhaps use its dominant position to extract particular privileges and rewards). In its most simple guise, power has therefore been written of as the story of how powerful people impose their ideas and values on the groups below. Over time, a number of different institutions and individuals – notably, the monarchy, the Church, the aristocracy, politicians – have presented themselves as the rightful holders of power and the ultimate arbiters of what is right and wrong, good and bad, legal and illegal. In medieval and classical times, for example, it has been noted that the authority of the sovereign to make these distinctions came, in large part, through the co-option of religious values and the claim that authority was 'god-given'. This legitimized sometimes despotic and autocratic actions to punish those who refused to recognize the monarch's values and rules: public floggings, executions, and so on. In more recent times, however, the (secular) state is recognized as the central institution responsible for disseminating ideas about right or wrong forms of behaviour, stipulating which sort of behaviours should be encouraged and which discouraged. These moral ideas are constructed through the bureaucratic rituals and traditions of political debate, with the creation of laws having become increasingly important for governing what we can and cannot do (Donzelot, 1979). Clearly, leaders of secular states do draw on religious beliefs (and are influenced by the religious histories of their countries) in their policy-making, and in the case of the UK, for example, the Church of England retains formal links with the process of governance.

Considering this process of state governance, Richard Sennett (1994) has argued that all societies need strong sanctions, rules and norms to make people aware of and conform to a social order that is based around certain values and interests. The way in which particular political groups have succeeded in imposing their values and rules on society is, however, far from straightforward, involving a complex process of domination, coercion and persuasion. Here, the ideas of the Italian neo-Marxist Antonio Gramsci are often regarded as helpful for understanding how power is acquired and wielded by ruling groups. Writing in the 1930s, at a time when Italy had come under the control of Mussolini's fascist party, Gramsci strove to comprehend why an essentially repressive and undemocratic party had acquired popular support. For him, only part of the answer was to do with the use of military force and physical intimidation. Instead, he began to think about the way in which power was acquired through **hegemony** – a concept used to describe the 'organization of consent' based on the acceptance that a particular set of ideas, beliefs and values were desirable, inevitable and 'natural'. Gramsci began to focus on the power of persuasion, and the way in which Mussolini had been able to use rallies, public meetings and the mass media as a means of creating a cult of fascism that seemed to suggest Italy was the cradle of civilization and that the spirit of the Roman Empire could be recaptured through adherence to a programme of state fascism. These ideas apparently legitimized Mussolini's political leadership, with Gramsci postulating that his ideas would only be overturned once the Italian people were able to develop counter-hegemonic ideas which challenged the taken-for-granted assumptions underpinning Mussolini's ideas.

Ideas of hegemony are useful for understanding how particular political groups are able to assert claims to govern (and tax) a particular territory. But it is important to realize that hegemony is an ongoing process whereby particular groups and individuals need continually to reassert their right to rule. One way in which they do this is by seeking to inscribe their values on specific places and times, celebrating and justifying their rule through symbolic rituals and performances of power. Here, we might perhaps consider the way that political power is celebrated in particular military parades, state ceremonies, national holidays and celebrations that serve to legitimize the authority of the state (for example, 4 July in the USA or Guy Fawkes' Day in the UK). Beyond this, we might also think about how the power of the state is symbolized in the construction of spectacular parliament buildings and national assemblies or the design of civic spaces and monuments (see also the discussion of national identity in Chapter 5). For example, in reviewing the importance of the streets in fascist Italy, Atkinson (1998) has noted that the hegemony of the Fascist Party was symbolically secured through a grand redesign of Rome that drove new boulevards through the heart of the city, providing new ceremonial spaces in the process. Less spectacularly, perhaps, at the heart of most towns and cities there are civic buildings and town halls which symbolize the authority of the (local) state through their use of classical or ornate architectural styles, incorporation of civic insignia and dominating façades (Miles, 1997). Statuary may also be important for celebrating certain senses of place, with public art often used to legitimate attempts to remake place identities (see Figure 8.1).

But this account of the workings of political power perhaps fails to acknowledge the way that the power of the state may be spread far more widely, into a variety of places in the centre or on the margins of society. Importantly, the transition from a form of state control in which a sovereign state governed space autocratically, to one where this control centred on the education and reform of 'marginal' citizens

Figure 8.1 Statuary as a performance of power: *Forward*, Centenary Square, Birmingham

Photo: Authors

(i.e. those who did not conform to standards of 'normality'), featured prominently in much of the work of Michel Foucault. Although Foucault's ideas about power and knowledge are not unproblematic (having been subject to critique from several angles), they have proved influential in many subsequent explorations by geographers and historians as to the relations of power, knowledge and place. As such, it is perhaps worth spending some time considering his arguments, as although they focused on a particular series of moments in the history of social relations, they have been taken as having wider relevance in the examination of power relations.

In essence, Foucault's account of the changing nature of social control focused in on the period where the medieval, feudal system was replaced by a quintessentially **Modern** social order in the late eighteenth and nineteenth centuries. For him, this transition to a Modern social order was made possible by the imposition of new forms of control dispersed through complex networks of knowledge and power. He illustrated this by focusing on the lives and experiences of those figures who lay on the margins of society – particularly convicts, the mentally ill and the poor – suggesting that the way their behaviours were controlled changed massively at this time. Specifically, Foucault reasoned that Modernity emerged from the culmination of developments in science and philosophy which collectively rejected the idea that order could be simply maintained through sporadic displays of bodily (or corporal) terror such as torture or public executions. In his book *Discipline and Punish* (1977), for example, he describes graphically the forms of public punishment that were used to impose autocratic state control, essentially by setting an example to other citizens. Foucault's descriptions of these punishment rituals are not for the faint-hearted, stressing that they were very much used to strike fear into people. He uses, for example, contemporary documentation about the public execution in 1757 of the French regicide (a person who murders a king) Damiens, which states that 'the flesh will be torn from his breasts, arms thighs and calves with red hot pincers, his right hand . . . burnt away with sulphur, and, on those places where the flesh will be torn away, poured molten lead, boiling oil, burning resin . . . and then his body drawn and quartered by four horses and his limbs and body consumed by fire . . .' (quoted in Foucault, 1977, 3) Here then, people were strongly encouraged to behave in particular ways, observe certain rules and seek to attain certain standards of living, knowing that if they did not match up to the state's idea of what was normal or good behaviour then their bodies were at risk of torture or death.

Leaving the premodern apparatus of the state, Foucault then turned to explore the forms of control evident in 'Modern' nineteenth-century society. Here he noted the emergence of seemingly more humane and democratic control of those whose behaviour didn't fit into the state's idea of what was normal. Criminals, for example, were imprisoned; the 'mad' were incarcerated in mental asylums and clinics; the unproductive were placed in 'poor houses'. In these spaces, they could be kept away from good citizens lest their behaviour influenced or disrupted others; moreover, they were subject to forms of regulation and inspection designed to reform and modify their behaviour. As such, these institutional settings served to organize, classify and control those characters who, in the language of the day, had been 'deformed through the pathologies of mental or physical disability' or 'the diseases of crime, drunkenness and sloth'. Such institutions were, therefore, places of considerable order and regularity; meals were taken at set hours, inmates might have set exercise periods; permitted pastimes were tightly constrained and visitors only admitted under certain circumstances. As **sites of reformation**, these buildings were designed to encourage the adoption of a simple, regular and productive life which could enhance the strength of the state, turning its inmates into 'useful members of society' (Foucault, 1988).

Therefore, although Foucault's wide-ranging writing referred only fleetingly to the way in which people and place are implicated in broader relations of power, his emphasis on the social order implicit in institutional spaces such as the asylum and the prison has been seized upon by geographers endeavouring to show that the relations between people, knowledge and power are transformed into actual relations of power in specific places. In particular, Foucault's use of the **panopticon** as a general metaphor for the forms of spatial arrangement which were central to the disciplining of deviant individuals and behaviours suggested to geographers that specific principles of **spatial segregation** and **surveillance** were essential in the maintenance of social order. This panopticon model was developed by the social reformer Jeremy Bentham in a series of letters and correspondence from 1787 onwards (its name derived from the Greek for 'all-seeing eye'). As Figure 8.2 shows, what Bentham advocated was a new form of design for prisons in which principles of observation were crucial. A central feature of this design was that it would consist of numerous single cells positioned on the radii of a circle, each facing inwards towards an inspector's lodge from which it would be possible to see the actions of every inhabitant of every cell (through its iron grill). The prisoners would not be able to see into the lodge, and would therefore not know whether they were under observation at any given moment. According to Bentham, the threat of continual observation was supposed to discourage misbehaviour, with the visibility of inmates maximized through their spatial separation. Bentham also believed that the separation of inmates would inhibit the contagion of bad thoughts, bad behaviours and disease.

The historical geographer Chris Philo (1987) has explored the evolution and subsequent adoption of this panoptic design, describing how its blueprint was tailored and refined to specific local circumstances, as well as acting as a basis for the design of key institutions of reform, including workhouses, prisons and hospitals. In fact, few institutions slavishly followed Bentham's panoptic design, some allowing inmates to see and hear one another, others allowing them to see their keepers. But despite the variation in architectural design and style, new types of institutional setting began to emerge from the late eighteenth century, each seeking to impart order and discipline by bringing the building, its controllers and its inmates together in regulated space and time (see also Markus, 1993). These new sites consequently acted as the harbingers of a new social order, as the idea that bad citizens could be reformed spread through Western societies. Foucault's writing meticulously

Figure 8.2 Bentham's design of the Panopticon

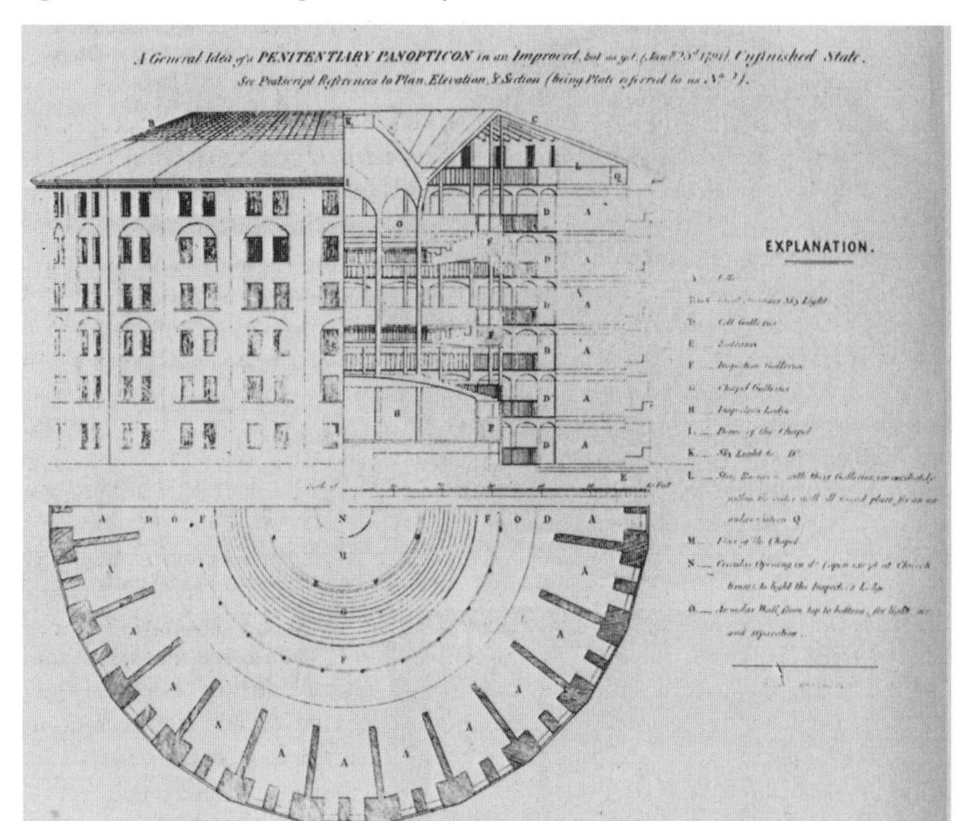

documented the way in which the local and central state dispersed these sites the length and breadth of the country so as to convey a general impression of social orderliness to the population at large. In a wider sense then, the creation of such disciplinary sites demonstrated to Foucault that an expanded, unified and intensified form of surveillance was being used to secure the boundaries of the state in the Modern era (see also Dandeker, 1990).

Because Foucault's academic exploration of the boundaries between sanity and madness mirrored his own experiences of drug-use and sexual experimentation, his writing was often controversially received. Yet it has proved extremely influential throughout the social sciences by offering a unique perspective on how social order and discipline can be imparted and instilled in place. Although preoccupied with the control and exclusion of those figures who exist on the margins of society, such as the mad, the sick, sexual deviants and criminals, his work nonetheless used these groups as a lens through which the workings of power in society as a whole could be more clearly viewed. For Foucault, the incarceration and treatment of miscreants in these sites was only part of the story of how state control came to be maintained through **disciplinary power**, which focused on the body and mind in such a way as to turn all citizens into 'good', docile citizens. More widely, he noted that the process of creating dedicated and productive citizens was institutionalized in state programmes of education, welfare and social policy designed to encourage people to aspire to 'normality', 'moral uprightness' and productivity. In this sense, disciplinary power may be thought of as **internalized** through people's individual self-control and adherence to shared moral codes.

Here then, Foucault's dissection of the way institutional managers sought to divide and rule good and bad groups can be extended into an understanding of the relationships being forged between people and places in other modern sites – such as factories, schools, universities, libraries and so on. Thomas Markus (1993) describes these institutional sites as places of **formation** where Modern ways of forming character were invented and imposed through specific power relationships, going on to talk about sites of **recreation** when describing the coffee houses, hotels, restaurants and swimming baths that were continually used by those privileged in these power relationships to redefine their own character and reaffirm their privilege. In many senses, the building types identified by Markus remain key sites of disciplinary power even today, with schools and colleges (in particular) representing sites where key ideas of 'wrong' and 'right' are instilled.

■ Exercise

In Chapter 3, the classroom was discussed as a behaviour setting. By the age of eighteen, most of us have already spent nearly 15,000 hours in classrooms of one type or another, but we are rarely encouraged to think about the design of these spaces. Thinking about a classroom or lecture hall you are familiar with, try to write down a list of the particular design features which might encourage particular forms of behaviour and understanding among students. Is there a particular form of power relation being played out here? How are techniques of segregation and surveillance employed in such classrooms or lecture halls?

In considering this question, you may have thought about a wide range of features on a variety of scales. On a microscale, for example, you might have thought about how the provision of chairs and desks encourages people to sit upright rather than slouching. There is also probably a set gap between desks to maintain some degree of separation between students, while the arrangement of seats and desks so, is generally designed to encourage a particular form of student participation, placing the students under the constant gaze of the teacher/lecturer (and vice versa). Of course, the layout of the room will also depend on the type of lessons that are supposed to be taught in it, reminding us that particular places are deemed appropriate for imparting particular knowledges (e.g. a laboratory for science, a studio for art and design, a computer room for information technology). Yet whatever type of subject is taught in the classroom, it is inevitably an ordered environment, having certain areas that are 'out of bounds' to students (except in specific circumstances). Ultimately, this demarcation between tutor's and pupil's space reinforces the type of power relationship that might exist in this setting.

Some writers have extended this analysis to consider places of education, like schools and universities, as sites where particular social orders are created and maintained through more diffuse rules, norms and practices. Writing of such sites as 'citizen-making machines', they have acknowledged that educational settings instil a particular relationship between students, knowledge and objects of study (books, laboratory equipment, computers, etc.), with the design of teaching rooms for various subjects reflecting a wider understanding of what constitutes appropriate knowledge, and setting limits on what can be taught where. The creation of a nationally approved curriculum may in turn reflect certain assumptions about what is 'useful' knowledge, with pupils taught particular sets of practical, social and personal skills according to ideas about what will be 'useful' in the 'wider world'. Moreover, school rules and dress-codes encourage conformity and participation in a way that is supposed to encourage each pupil to become a responsible individual by subsuming his/her individuality in the interests of the wider school community. Here, a focus on the pupil's mind is matched by a focus on their bodies with cleanliness, healthiness and athleticism encouraged through periods set aside for physical education, team sports, sex education and even medical inspections. Viewed from this Foucauldian perspective, the school becomes a place of constant surveillance where undesirable forms of behaviour, morality and appearance are discouraged through coercion and punishment, while good behaviour and learning is rewarded by the granting of privileges, good marks and praise from tutors.

Hence, much of the geographic research inspired by Foucault's work has explicitly examined the 'technologies of domination' and power used to normalize behaviours in enclosed (or **carceral**) spaces, particularly those institutional settings which are designed to produce healthy, respectable and civilized individuals (see Markus, 1993; Ogborn and Philo, 1994; Ogborn, 1998). Here the focus is on both mind and body, with the elimination of 'distasteful' bodies seemingly as important as eradicating 'abnormal' acts. However, it has recently been argued that some geographers have taken Foucault too literally in focusing on Bentham's panopticon and its design variants, with Hannah (1997, 344) stressing that there are two principal limitations evident in this writing: 'First, it has tended to remain pre-occupied almost entirely with the exercise of power in institutional settings; and second, to the extent that it does venture beyond institutional walls, it employs the image of the panopticon more as a motif or mood setter . . . than as a source of particular analytical tools'.

As Hannah argues, the disciplining mechanisms of surveillance are present beyond the walls of carceral settings, with all individuals living in spaces that subject them to a process of social ordering through surveillance – or more correctly, the threat of surveillance. After all, as Foucault himself was keen to stress, no spaces exist outside of power relations.

Foucault's idea that power is dispersed through a variety of everyday spaces and institutions should therefore alert us to the fact that all places are potentially implicated in the reproduction of social orders. In this way, we might be able to think about how a wide variety of places are characterized by power relationships that are based on control and discipline. This includes not only private sites (institutional sites, the home and the workplace) but also public and quasi-public places. For instance, recent debates about Closed Circuit Television (CCTV) on the streets of Western cities have drawn comparisons with Bentham's planned panopticon (Fyfe and Bannister, 1996). Like the panopticon, the CCTV schemes found in many Western cities meet Bentham's principle that power should be visible and unverifiable (i.e. that people should be constantly aware that they are being surveyed by some higher authority). In this sense, the presence of CCTV cameras on the streets is widely advertised, with many of the cameras being highly visible. Yet, at one and the same time, it is never clear to anyone in the city whether they are being watched at any particular moment. Most of us are only vaguely aware of whether our everyday movements are being observed, and would probably be very surprised to find how often we are within sight of a CCTV camera. For instance, one journalist based in London calculated that on his hour-long journey to work, he could be observed on up to 500 different cameras monitoring his passage from house to station and on to his final destination (Graham *et al.*, 1996). In effect then, almost every moment of our daily routines through the landscapes of towns and cities may be under the gaze of CCTV cameras, with private security cameras inside shops, offices, sports centres, banks, buses, trains and restaurants (as well as many residential areas) meaning that people leave a continuous trail of images on CCTV cameras in their daily lives (see Figure 8.3). In Foucauldian terms, this intensive surveillance is supposed to deter deviant behaviour by inducing a fear of permanent visibility. Matched by the threat of intervention by police, security guards or other forces of law and order, this permanent gaze on to the streets (in effect) is designed to assure the continuous 'automatic functioning of power' (Foucault, 1977, 201).

Justified as a cheap means of preventing crime, the estimated 150,000 professionally installed cameras in British cities have been widely praised as enhancing people's safety and regenerating many crime-ridden areas. But such claims rest on limited and sometimes dubious empirical evidence; moreover, although most citizens seem to welcome the presence of cameras on the streets (Graham *et al.*, 1996), there are dissenting voices who point out the potential for this technology to be abused. Depicted in some quarters as an infringement of civil liberties, it has been suggested that cameras are a threat to the democratic rights of citizens to go about

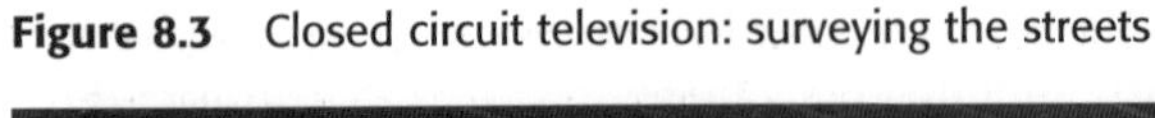

Figure 8.3 Closed circuit television: surveying the streets

Photo: © Clive Shirley/Impact Photos

their 'private' daily business. In addition, there are concerns that these cameras may be abused by the (largely unregulated) individuals and agencies who operate them. Here, it has been suggested that operators may 'over-scrutinize' those individuals whose appearance or conduct does not fit in with the dominant construction of what is 'normal' or desirable. Indeed, some evidence points to the way operators may use CCTV to expel 'troublesome others' – street vendors, the homeless, buskers and so on – from certain spaces. This is particularly the case in shopping malls where groups and activities seen as having no commercial value are excluded. Effectively summarizing much of this literature, Sibley (1995) thus concludes that various individuals, distinguished by their behaviour and appearance, are being expelled from these privatized spaces by other groups who feel that their appearance and conduct does not conform to the well-ordered ambience of 'family' consumption (see also Aitken, 1998).

For many geographers, therefore, the new technologies employed by the forces of law and order in the urban West, particularly CCTV, demand consideration in terms of their capacity for social and spatial control as much as their capacity for reducing crime rates (Fyfe and Bannister, 1996). Imbued dramatically with Orwellian overtones of 'Big Brother' by those seeking to find echoes of Bentham's panopticon in the contemporary city, these 'eyes on the street' are, of course, not evenly distributed, being strategically distributed to control the types of individuals and social groups present in particular 'public' places. Indeed, the idea that public places are becoming increasingly structured in favour of a narrowly defined social order is one that has preoccupied geographers and urbanists in recent years. Starting from the normative position that city life entails a 'being together of strangers' (Young, 1990, 237), it has often been argued that public space should be democratic and open, offering a site where difference can be acknowledged and celebrated. Zukin (1995, 260), for example, argues that public space is the primary site of 'public culture', where people may mingle and socialize in such a way that a shared 'citizenship' is created – a sense of urban identity transcending class, gender and ethnic boundaries. This idealization of public space suggests that the spaces of the street, the park and the city square represent truly democratic and open spaces where all may gather, free from exclusionary violence (Mitchell, 1997). As such, drawing on the arguments of Habermas (1991), public spaces have often been depicted as democratic sites of critical public discourse, of reasoned argument, and, on occasion, of protest and resistance (see Chapter 9). You might like to compare your own experience of such spaces with such an idealized perspective. Are they always sites of mutual tolerance and reasoned debate?

While there are important differences between CCTV in public spaces and the way power operates in private spaces, both are part of the elaborate technology employed by the state to produce docile, productive citizens. It is for such reasons that commentators like Davis (1991) have identified the recent emergence of new carceral spaces like the 'gated' residential enclaves and privatized shopping malls of Los Angeles as symptomatic of the death of public space in both a metaphorical and material sense. The seeming replacement of open and democratic spaces by closed 'sanitized' spaces in which certain groups and individuals are seen to be 'out of place' has therefore been described by Davis and others as representing a major threat to notions of citizenship defined in terms of equal access and participation in society. This issue even reached the High Courts in America, where protests over the privatization of public malls provoked Supreme Court Justice Marshall to observe that 'As governments rely on private enterprise, public property decreases in favour of privately-owned property. It becomes harder and harder for citizens to find means to communicate with other citizens. Soon, only the wealthy may find effective communication possible' (cited in Kowinski, 1982, 35). In such carefully planned mall environments, strategies of surveillance ensure that the atmosphere of conspicuous consumption and style is not broken by those who don't appear very interested in spending money. Accordingly, many analysts of the nature of contemporary public space are switching their focus to the way that some groups are being excluded from specific urban areas of consumption.

For example, the observation that public space is permeated by exclusions made on the basis of gender is one that has been remarked on extensively, with many feminist researchers having drawn attention to the importance of the private/public dichotomy in shaping women's lives (Duncan, 1996). The origins of this **gendering** of public and private space can essentially be traced back to the beginning of the Modern period, when new ideas about power and knowledge were defined primarily in bourgeois masculine spaces such as the coffee house, tavern, park and garden, rather than the private, domestic sphere:

> Whereas in the eighteenth century, it is claimed, the home tended to be above the workplace in the city and women might be involved in both the commercial and domestic realms, by the beginning of the nineteenth century the bourgeois home tended to be separated from the workplace with the paterfamilias [male head of the family] travelling between the home and work in the city and women carrying out their roles as wife/mother/homemaker in the domestic sphere.
>
> (Nead, 1997, 659)

The development of the ordered suburbs increased the isolation of domestic life from economic production by isolating women not just inside the home, but inside a community far removed from offices, factories and civic spaces. Feminist geographers and historians have thus emphasized how, from the first moments of Modern suburban development, the distinction between city and suburb was imbued with ideas about separate spheres for men and women in which the public domain of the urban centre was both deeply masculine and associated with social, economic and political power. In contrast, the suburb came to be associated with middle-class domesticity, femininity and dependence, with the cult of domestic interior design fixing feminine bourgeois values in a space where women could submit to a fantasy that was unrealizable in the exclusionary spaces of the masculine city (see Chapter 6 for further discussion of the ways in which the city and suburbs have been imagined).

In fact, the division between public and private is much more complicated than the simple binary division implied above (i.e. male public spaces and female private spaces). Instead, there are very complex relationships between gender, place and power, as Linda McDowell (1999) demonstrates. But the idea that men and women enjoy **unequal access** to particular places is a very important insight generated by feminist geographers. As has been discussed above, powerful agents within society construct space in a way which reflects the structure of power in society, and this is as true for gender as for the differences between rich and poor, black and white, young and old, homed and homeless (and so on). Therefore, although the distinction between private and public space has been consistently challenged and recast, it remains a potent means by which women are controlled, excluded and oppressed (Bell, 1995; Duncan, 1996). Assumptions about where women and men belong may be challenged and transgressed, but in general it appears that women are still excluded from some places and made welcome in others in ways that perpetuate the dominance of masculine over feminine values in society (referred to as **patriarchy**). We shall return to debates about how gender and sexual roles are shaped in different places in Chapter 9, when we consider the relations of sexuality and space.

■ Reading

The chapter 'Taking charge in modernity' in Shurmer-Smith and Hannam (1994) offers a useful overview of the literature discussed in this section, with Ogborn and Philo (1994) offering an interesting historical example of how Victorian institutions in Portsmouth (UK) were supposed to impart social order. On CCTV and social exclusion, see the various essays in Gold and Revill (2000) focusing on the urban and rural 'landscapes of defence' which have recently emerged in Western nations. Although we have only touched very briefly on the role of the police as agents of social control, Herbert (1997) offers an interesting insight into the way they act to enforce the law in different situations.

■ 8.3 Civilized bodies, civilized places

An appreciation of how the gaze of the state and law penetrates into the nooks and crannies of everyday life is indicative of the way in which we are caught up in the networks of regulation that shape the relationships we have with the places we inhabit and occupy. However, it is useful – and perhaps essential – to look beyond these overt expressions of state control to examine the processes of regulation and 'normalization' that lie outside the formal processes of the law and overt state control. Invoking Foucault's work on the diffusion of state power into the 'sites, thresholds and screens that may lie hidden or silent in the everyday practices of civil and social life', Fischer and Poland (1998, 187) contend that it is too easy to imagine state control merely in terms of 'the visible arm of sovereign power' manifest in military might or the forces of law and punishment. Indeed, while the focus on the role of legal and governmental institutions and organizations in shaping geographies of everyday behaviour can, perhaps, indicate the 'limits' of acceptable and antisocial behaviour in a particular society, such a focus downplays the wider and more diffuse relationship which exists between the state and **civil society**.

Here, the term civil society is used in a specific sociological sense to describe the social relationships and interactions which occur outside the sphere of the state (Tester, 1992). According to many sociologists, the reproduction of society is essentially organized around the day-to-day relationships which are apparently uninfluenced by the state, yet which are still characterized by gender, ethnic and class inequalities. These relationships are essentially ones driven by self-interest, with social bonds and attachments constructed between people in civil society on the basis of emotional attachments and attractions centred on desire, disgust, aspiration and revulsion (Lupton, 1998). These emotional ties and relationships are inevitably influenced by social rules and norms, with certain behaviours, bodies and modes of acting becoming privileged over others, resulting in a distinctive social hierarchy. This social hierarchy is manifest in (and maintained by) everyday activities like sitting, eating, talking, drinking and washing, as well as in ways of walking, modes of speaking, and styles of dressing. Obviously the state has only a limited amount of control over these facets of daily life (e.g. few styles of clothing are illegal, even if we think they are fashion crimes!), but the way that people aspire to or reject dominant ideas about behaviour in the civil sphere is crucial in attaining social standing or success.

Here, the French sociologist Pierre Bourdieu has developed ideas about social standing and cultural capital which begin to explain why such behaviours (or 'performances' of the self) are crucial for defining the social relationships we share with others, as summarized by Bryan Turner:

> Social status involves practices which emphasise and exhibit cultural distinctions and differences which are a crucial feature of all social stratification . . . status may be conceptualised as lifestyle; that is the totality of cultural practices such as dress, speech, outlook and bodily dispositions . . . . While status is about political entitlement and legal location within civil society, it also involves, and to a certain extent is, style.
>
> (Turner, 1988, 3)

The idea of distinction then is closely connected with **style**, not merely in the sense of fashion, but in terms of what type of people, practices and places are culturally (and commercially) valued. A common cliché is that matters of taste and style are down to the individual, but in fact it is obvious that there may be many shared ideas about what is tasteful or stylish in a given context. As such, certain ways of dressing, speaking and acting may be generally considered as stylish, hip and desirable, while others may be considered as common, uncouth or naff.

While policy-makers and governments have little influence over shaping these judgements, it is apparent that some social groups have more power than others to determine what is desirable and where. Fisher and Poland (1998) provide an interesting example of this when they examine changing perceptions of smoking in public places. Issues relating to smoking have long been the focus of state projects to instil certain moral and physical values as desirable, and in postwar years at least, there has been a concerted effort to point out the dangers of cigarette-smoking in many Western nations. Yet simultaneously, the state's enthusiasm to eliminate smoking has often been tempered by liberal notions that people should have a free choice in terms of what they do with their own bodies (perhaps also mindful that the tobacco industry is an important source of tax revenue). Instead, it is primarily private businesses and organizations which are introducing bans on cigarette-smoking in a number of workplaces, shops and restaurants. Such decisions to ban smoking are often justified with reference to the 'rights' of non-smokers to breathe smoke-free air, but an underlying concern seems to be that smoking is now regarded as a socially undesirable activity, one which 'lowers the tone' of a particular place. Of course, most organizations or companies that ban smoking do not have any legal power to enforce this ban, relying on the smokers to regulate their own behaviour through respect for others (observing social etiquette). Nonetheless, the public places left available to smokers symbolize that smoking has now generally been 'squeezed out' from valued places: smoking is largely restricted to outside office buildings or in a shrinking number of bars and restaurants. As Fisher and Poland (1998) note, the representation of smokers as 'outsiders' in contemporary society is neatly mirrored in the exclusion of smoking to outdoor spaces.

Debates about smoking therefore provide a useful illustration of the way that power relations are played out through interactions of the state and civil society. A seemingly less contentious example of a behaviour that might be seen as 'out of place' is the consumption of food in public spaces. In Western societies, the preferred setting for eating has often been defined as the private realm, where the act of sitting down to eat at the table (using appropriate cutlery for different foods) has been suggested as distinguishing human eating practices from those of animals. Yet millions of people ignore this everyday by eating while walking through the streets, at the wheel of their car, sitting on a park bench and so on. According to Gill Valentine (1998), this type of behaviour has connotations of animalism, primitiveness and wildness. Moreover, with the hands used for consuming food (rather than cutlery) there may be more risk of food dropping from the grasp, sauces dribbling down one's face or crumbs adorning one's clothes. As such, those who eat in this way may be viewed as clumsy or unsophisticated, their personal comportment and status brought into doubt because of the way they 'choose' to eat. Indeed, as one of us found when exploring local people's responses to tourists in Stratford-upon-Avon (a popular destination for day-trippers because of its association with William Shakespeare) one of the major bones of contention for some local people is that tourists are spoiling the town's ambience by consuming fast food in the streets and parks (see Figure 8.4). Here, the local residents were expressing very real concerns about litter polluting 'their' streets, but it is possible to discern an underlying fear here that the sight of people eating with their hands in some of the town's most beautiful parks and open spaces was 'lowering the tone'. Similarly, in other towns, bans on the consumption of alcohol in streets and parks have been introduced – ostensibly because of fears about public order – but have been used to target people drinking beer in the street rather than 'families' or 'couples' enjoying a glass of wine in a public park (Oc and Tiesdell, 1997). Indeed, in cities like Coventry where an alcohol ban has been in place since the late 1980s, it has been only selectively applied so as to facilitate the development of a 'continental' café culture (showing that public consumption of food and drink may take on very different meanings according to where and how it occurs).

Analysis of how we consume food in public and private again highlights the complex workings of power in contemporary society. Indeed, although the consumption of food may be controlled by regulations and laws in some places (with many geography departments we know prohibiting staff and students from eating in the corridors because it looks 'unprofessional'), more generally this is a matter for self-restraint: 'There is no need for arms, physical violence, material constraints. Just

Figure 8.4 Impolite behaviour? Eating on the streets, Stratford-upon-Avon

Photo: Authors

a gaze. An inspecting gaze, a gaze which every individual under its weight will end by interiorising to the point that he [*sic*] is his own overseer, each individual exercising surveillance over and against himself' (Foucault, 1977, 155). As Valentine (1998) writes, this fear of the gaze – of being seen to eat in public, has put a brake on the pleasures of street food for many potential consumers. Like many other activities that many people routinely participate in (like drinking, smoking, defecating, washing or having sex), eating is thus regarded as belonging in the private, domestic realm or in specially dedicated places like restaurants. The reasons for this are complex and varied, and may be traced to deeply ingrained anxieties about the maintenance of our self-identities. As Douglas (1966) has asserted, solids, liquids and gases that cross the boundaries of our body (literally passing from outside to inside, or vice versa) seem to problematize our definitions of who we are and where we end, with the ingestion or expulsion of matter from the body surrounded by a whole series of rituals of cleansing (see also Chapter 4). For example, we demand our food is clean before we eat it, purify drinking water, wash our hands after urinating or defecating and use deodorant to stop sweating. In this respect, power (in all its various manifestations) seems to operate not just to discourage or eliminate distasteful acts and practices, but to discourage distasteful **bodies** too. Good, complete citizens are generally considered to be those who look after their bodies, maintaining and policing its boundaries routinely and rigorously.

The idea that our individual bodies are caught up in power relations offers an interesting insight into the relationships that exist between people and place. Commonly, we take our bodies very much for granted, the product of our biological makeup. As 'naturally' given, the body may be seen as relatively fixed and unchanging, providing us with some certainty about our physical identity. Against this, it is obvious that the decoration, display and alteration of our bodies has been a major concern since ancient times, whether through tattooing, piercing, plastic surgery, aerobics and body-building (Figure 8.5). This means that the body is not just biologically given, but socially constructed, in the sense that our bodies take on different meanings according to shared sets of meanings which categorize and differentiate between people. Bodies are not just physical, fleshy objects – they connote social identity. Different body sizes, colours and shapes are therefore read as indicating something about the individual who possesses and moulds that body – that they are a man or a women, black or white, straight or gay, upper-class or lower-class, healthy or unhealthy, desirable or undesirable. Consequently, these meanings exert a profound influence over how individuals seek to manage or alter their bodies so as to influence the way others react to them. Clearly, individuals will wish to present themselves differently in different places – dressing for the office and dressing for a nightclub means trying to impress different people in very different situations. As such, the body may be seen as a product of socialization but also as a dynamic project of the self as it is made and remade in everyday life (Lupton, 1998). The body, therefore, is a key mediator between self-identity and social identity, literally taking its place in the world as the physical expression of individual identity.

Examining the history of the body, a number of sociologists and psychologists have noted the way that the meanings associated with different bodies have changed over time according to different notions about what types of bodies were most desirable. For instance, Shilling (1993) has noted the influence of different classificatory systems in determining the type of 'body projects' which people enter into (i.e. the way individuals work at their body identity). In so doing, a theory of the **civilized body** is outlined, whereby the body has been transformed over the centuries into a polite and sophisticated object to be possessed and managed by its 'owner' in opposition to everything defined as animal or associated with 'wilderness' (see Chapter 6). In contemporary Western societies, the civilized body is one essentially disciplined at the individual level, with boundaries between different individuals only normally breaking down in the context of romantic (and private) relations with a partner. Here, behaviours which might be unacceptable elsewhere (e.g. passing food from mouth to mouth, feeding one another) take on different, perhaps sexual, meanings.

Yet this focus on the individual civilizing her or his body was not always the case, and Lupton (1998, 72) has noted the way that urges to exclude certain 'deviants' from medieval cities were connected to specific understandings of the body. She contends that the body was imagined at this time not as a private body but very much

Figure 8.5 Working the body

Photo: C. Rowley

as a public body, in that acts later regarded as distinctly private (such as defecating, eating, farting and so on) were commonplace in public. This reflected a wider understanding that the body was a porous container of fluids, humours and blood, with a constant and 'natural' exchange between inside and outside. In this way, people were encouraged to see their bodies (and, consequently, their selves) as part of the wider community in which they were living: problems of pollution and disease were not so much failures of individual lack of self-control but failures of communal order. In this way, the boundaries between self/other were defined in collective terms rather than being negotiated at an individual level, and those regarded as having diseased, grotesque or polluted bodies were removed from the social body in a fairly clear-cut manner. Lepers, prostitutes and sometimes Jews all suffered because of this, their bodies seen as the source of pollution and contagion (Sennett, 1994). Here, the authority of the Church was foremost in drawing the boundaries between appropriate comportment and 'bad' behaviour, disciplining and purifying people's bodies at a collective not individual level.

Descriptions of how conceptions of the body have changed over time serve to stress that bodies, often thought of as natural, are inextricably related to and bound up with social relations. Today, then, bodies (which are of course inseparable from minds) are frequently focused upon by geographers intent on exploring the relationships that exist between people and place. As such, examining what our bodies mean – and what we can do with them – in different places offers an interesting perspective on the geographies of everyday life. The edited collections by Pile and Nast (1998) and Teather (1999), for example, have addressed the connections between people and place by exploring the complex way that bodily identities are performed and made sense of in different places – in the home, at work and at play. These identify some of the ways in which particular types of body are read as being male or female, straight or gay, white or coloured, healthy or sick, and, consequently, how certain types of body are excluded from certain places (again stressing that certain institutions and places are responsible for encouraging the production of civilized, idealized bodies). As was noted earlier, schools may be implicated in encouraging self-management of bodily performance through the imposition of rules and regulations relating to dress, bodily adornment or body-piercing. Likewise, sports settings are key sites for the (often theatrical) display of athletic, toned bodies (Eichberg, 1998). As a result, uncivilized bodily performances may be considered threatening in these contexts, with discrepancies from acceptable appearance codes prompting punishment or dismissal. 'Gender-bending', where men or women change their 'natural' gendered bodies by cross-dressing, make-up or plastic surgery, may especially provoke anxiety in many social settings because of the way it transgresses expectations of what 'straight' female and male bodies should look like (see Chapter 9). Similarly, women breast-feeding in public may experience negative reactions because of the transgression into public spaces of what is frequently considered to be a private activity, and because of the dual significance of the breast as a sexual object and a way of feeding babies.

An interesting case-study of how some bodies may be read as 'out of place' is provided by Robyn Longhurst (1998) in her examination of the pregnant body. Interviewing pregnant women in Hamilton, New Zealand, she found that the Centre Park shopping centre – a shopping environment targeting middle- and high-income women – was experienced as an exclusionary space. This sense of exclusion was evident on a number of levels; for example, it was evident that window displays often incorporated idealized images of women that are normatively glamorous, sexy and attractive, but that pregnant women did not fit into this conception. Moreover, few shops provided clothing for pregnant mothers. More prosaically, those interviewed also reported that toilets, escalators and seating were not designed with the bodily form of pregnant women in mind. Concluding, Longhurst stresses that there are many social norms which pregnant women are supposed to adhere to in terms of dress-code, comportment and movement in public space. Because of cultural rules of pregnancy whereby women are not supposed to engage in sport, drink, smoke, or have sex, it is not surprising that some women find their usual public behaviours becoming unacceptable. In short, women whose bodies are sexed (and sexy) before they are pregnant may find that they take on different meanings during pregnancy, meaning that they may feel 'ugly and alien' in some places that they previously felt welcome in.

Of course, we need to be careful about generalizing about the meaning of bodies in different contexts (cf. Seamon, 1979 – see Chapter 4). Certainly, desirable body forms vary across time and space in important ways. For instance, the obsession with body image in Western societies today is evident in prominent debates over obesity and slenderness. Throughout history, most Western societies have repeatedly demonstrated a social preference for body-types which are neither too thin (and thus lacking in muscle mass and strength) nor too fat (and whose bulk inhibits the ability of the body to move). During this century, however, females have come to diverge from males by preferring a thin or **ectomorphic** body-type for themselves, with thinness being interpreted as a sign of high social status (and power). The media have long promoted the ectomorphic body-type as the ideal shape for women, while in general images of men tend to depict them as more massive and muscular.

■ Exercise

Collect a range of life-style, fashion and health magazines and conduct a basic content analysis of the type of images of masculine and feminine body forms which they incorporate. This could involve going through a sample of these magazines and noting the number of adverts which include men or women in different poses, activities, and states of undress. Think about the following:

- How many adverts show women's bodies relative to men's?
- Do men's bodies take up more space than women's?

- Are women's bodies portrayed as subordinate to men's?
- How many images display obese bodies (and why)?
- Do the images of men's and women's bodies differ according to the magazine's potential readership?
- Do idealized body images for men and women differ with age, race or class?

The image of the thin, youthful, white (but tanned) woman is obviously not the only one you will encounter in the media, yet it has been one that has been widely idealized and commodified (literally, made into something that can be bought and sold). Predominantly then, this idealized slender, non-flabby, feminine body form is one that has encouraged women to display their bodies so long as they match this cultural idealization. For Susan Bordo (1993) the slender body has become valued as it expresses an ideal of a **well-managed**, disciplined self. Through watchfulness over diet and food-intake, sensible exercise and pursuing a healthy lifestyle, the ideal body is seemingly available to all women. Of course, anxieties about matching up to this body image are widespread, with compulsive disorders like anorexia nervosa seen to result from pressures to diet in a consumer society where food is 'sold' as central to many social rituals. Equally, anxieties about obesity as representing a failure of self-management are also widely apparent. Here, sexist assumptions about women's rights to display their bodies in particular spaces may serve to discourage women who consider themselves 'overweight' from sunbathing in public parks or beaches, for instance. Equally, returning to Valentine's (1998) discussion about food consumption, many obese people feel extremely uneasy about eating in public space for fear that their 'uncivilized eating' may be read as evidence of their gluttony. Concerns about sex, food and health, it seems, are therefore central in determining the types of relationships we have with both ourselves and others in contemporary society.

■ Reading

Bell and Valentine's (1997) *Consuming Geographies: You Are Where You Eat* offers an enjoyable romp through the foodscapes of the urban West and provides some interesting insights into how the consumption of food is shaped by social conventions of about where we ought to eat what. The collection edited by Fyfe (1998) offers a more wide-ranging but equally accessible overview of how the streets of Western cities have become implicated in the dynamics of social exclusion, with even acts like smoking, eating and drinking prompting debates about appropriate behaviour in public space. In contrast, some debates on bodies and performance are difficult to get into; Butler (1999) offers a very accessible introduction, with Shilling (1993) providing a useful overview of connections between body forms and civilization.

8.4 Place and moral order

Considerations of which bodily performances are acceptable where have spurred many geographers to engage with wider issues of moral order. The term **moral geographies** is one that has emerged in recent years to describe empirical research into those aspects of socio-spatial ordering which 'invite a moral reading' (Smith, 1997a, 587). As such, an increasing body of geographical research has investigated the judgements which people make on an everyday basis about what type of people, behaviours and embodied practices are acceptable or appropriate in which settings. Developing such themes, Matless (1995) has argued that moral geographies operate not only through powers of disciplinary control and exclusion, but also through the powers of individual spatial practice. Taking various texts as their subject matter, many geographers have sought to detect a moral geography of taste and manners that encompasses a series of cultural expectations of what should happen where. Here, attempts to exclude certain activities from particular spaces have been examined against the backdrop of wider social debates about what constitutes moral behaviour. Returning to one of the aforementioned examples, attempts to prevent people from smoking in certain public spaces can only be fully understood by examining now widespread ideas that only weak-willed, negligent and unintelligent people choose to smoke. As such, the sight of an individual smoking is one that seemingly indicates that this person has values which do not accord with those of upstanding, moral members of society.

Matless (1995) has consequently identified a coalescence of the 'social' and 'cultural' around the theme of moral geographies, arguing that by linking the social geography of the country and the city to wider frames of reference (specifically the existence of moralizing discourses about the appropriateness of different types of behaviour) research has generated new insights as to how space contributes to social order (see for example, Ogborn and Philo, 1994). This type of study has a much longer precedent, with notions of morality explicit in the urban sociology pioneered by the so-called Chicago School in the early part of the twentieth century: 'Investigation might well begin with the study and comparison of the characteristic types of social organisation which exist in . . . bohemia, the half-world, the red-light district and other moral regions pronounced in character' (Park, 1916, 602).

The places particularly associated with alternative senses of morality can therefore be interpreted as betraying wider moral orders. The fact that disadvantaged and run-down areas are often associated with prostitution, for example, reminds us that selling sex is seen as an immoral act in contemporary society. From such a perspective, geography – in the sense not just of simple location, but of where acts are deemed appropriately to belong – is crucial to any understanding of power relations (Matless, 1995).

Yet as the consideration of public eating suggested, it isn't simply the sight of certain acts that prompts moral approbation (or approval), but also the **smell**. Again, this is related to systems of cultural distinction, with the pungent smell of fish and chips or a take-away kebab regarded by many as much less desirable than (say) 'exotic' or 'organic' foods. This is even played on by superstores and cafés who may aim to create particular smellscapes which encourage consumers to make particular purchases. One example of this you may have encountered is the way that many superstore chains waft the smells generated by the in-store bakeries throughout the store (see also Chapter 3). The consumption of smell can also become a key ingredient of the act of dining out. As Finkelstein (1989) writes, when people eat out they are pursuing a variety of needs and desires; going to a restaurant is about more than just hunger; it is about a sense of excitement and a sense of self-fulfilment which is supposed to derive from conspicuous consumption. The decoration, architecture and packaging of restaurants is a crucial aspect of this, and being offered new sights and smells is very much part of the dining-out experience. In some restaurants, the kitchens may be opened up to the public (or chefs may cook at the table) so that diners can appreciate the exotic ingredients and smells. Lupton (1998) thus writes that smells can produce emotional states; certainly, the smell and taste of coffee can invigorate and get one ready for the day ahead, while perfumes worn by a lover may be sexually arousing. On the contrary, other smells may repulse, with body odour often considered embarrassing in Western societies and the smell of smoke often interpreted as a **stigma symbol** (Fischer and Poland, 1998). Equally, residents living near football stadia in the UK often complain about the odour of hot-dog stands on match days, with the smell of frying onions lingering long afterwards thanks to the unfortunate tendency of some fans to discard napkins and half-eaten purchases into residents' gardens.

The idea that there is a geography of smell is not one that has been subject to much scrutiny by geographers. Indeed, Pocock (1993) notes that geographers have done little more than cite its evident role in arousal, warning and memory, with our knowledge of space being based on sense of smell as well as vision (see Chapter 3). However, some clues as to what a geography of smell might be like can be discerned in work conducted by a number of social historians writing on public health, sanitation and smellscapes. Notably, Steven Halliday's (1999) *The Great Stink of London* offers a wonderful overview (or *oversmell*) of the smellscapes of nineteenth-century London, while Alain Corbin (1994) considers the changing odours of Paris in *The Foul and the Fragrant*. This charts a story of the technological and medical advances that defined unwanted personal odours and then sought to eliminate them from the private and public realm. In doing so, he notes that smells have had different connotations across space and time; body odour was once considered a sexual stimulant, becoming regarded as unpleasant only in more recent times. From the end of the eighteenth century, in particular, it was regarded as appropriate for men and women to be clean, pure-smelling and wholesome. As Markus (1993) also asserts, this idea was inculcated through institutions such as schools, prisons and workhouses where strict regimes of cleanliness were imposed through the (enforced) washing of bodies and clothes. Outside institutions, it was encouraged through the provision of public baths and wash-houses, while in the home, the concern with eliminating odours and dirt was evident in the separation of rooms according to bodily functions, so that smells associated with cooking, washing and defecating would not mix and mingle with one another.

Perhaps the most obvious attempt to eliminate such odours from public space was in the attempts by city engineers to introduce new municipal services designed to sanitize urban space (Illich, 1986; Ratcliffe, 1990). This was accompanied by new experiments with chemicals and disinfectants designed to combat (in particular) the smell of faeces. This included attempts to mask the odours which lingered around cesspools by substituting alternative, more pleasant smells, while absorbants like charcoal and lignite were used to soak up the odours. Later, the use of oxidizers was to prove more effective as a means of disinfecting cesspools. Yet ironically, the increasing popularity of washouses and baths added to the problems of disposing of sewage, and stressed the need for a more sophisticated sewerage system. As Corbin (1994) notes, the installation of a sewerage system in cities like Paris became a major preoccupation of elite groups in the nineteenth century as they sought to impose their middle-class morality on to the geography of the city.

Like other interventions in the nineteenth-century city, the introduction of street-cleansing, sewage systems and refuse removal services was informed by particular understandings of links between environment and morality. Felix Driver's work on the relation of Victorian social science and environmentalism is interesting in this regard, demonstrating that nineteenth-century sanitary and medical science was predicated on a specific 'mapping of types of behaviour to types of environment' (Driver, 1988, 41). In so doing, Driver suggested that immoral values are often seen as being explicitly linked to physical pollution, with filth, criminality and disease conflated in the reforming rhetoric of nineteenth-century journalists and politicians. The city sewerage system consequently became a very real focus for both the hygienist movements prevalent at the time as well as an integral part of city-planning. The desire to eliminate bad smells thus had complex social and cultural foundations with the obsessive desire amongst many bourgeois individuals to rid the cities of certain smellscapes indicative of their concern about the immoral behaviour of working-class populations (see Figure 8.6).

The concern with smell was, and remains, shaped by social relations, with different classes claiming different sensitivities to smell. As such, smell may communicate social distance, with discourses relating to bad and putrid odours used to depict 'other' groups as undesirable or unacceptable in particular places. This argument has been developed in Sibley's consideration of **geographies of exclusion** (see Chapter 5), which offers one of the most thorough and theoretically sophisticated explorations by a geographer as to how urges to marginalize particular social groups feed off ideas of difference. Following Douglas's arguments about the maintenance of bodily boundaries, Sibley argues that the urge to remove bad smells from one's proximity is connected to deeply engrained and often subconscious desires to maintain cleanliness and purity, many of which may be inculcated in early infancy. The idea that powerful social groups may seek to

Figure 8.6 The personification of pollution: 'Faraday gives his card to Father Thames'

Source: Punch (21 July 1855) © Punch Limited

distance themselves from those individuals that they associate with 'dirty' smells, is an interesting one, though as yet largely unexplored in geographical writing (but see Cresswell, 1996).

But if smell has been neglected within geographical debates about the relationship between people and place, sound has been equally absent. Here, it is important to recognize a difference between type and volume of sounds. Though hearing ability differs significantly between different individuals, Rodaway (1994) notes that we may encounter both background and foreground sounds in our daily interactions with the environment. Background sounds are those that are largely constant or ambient, such as the hum of traffic in a city, while foreground noises are those that punctuate this soundscape. Like smell, sounds also evoke feelings and emotions (e.g. terror, fear, calm, peace, excitement and so on) and may therefore also be shaped by a complex moral geography. Debates about the appropriateness of sounds in particular locations normally focus on those foreground sounds which emanate from an individual source. For example, although constant traffic noise in a city centre may be annoying, few would probably complain about it, accepting it as a 'natural' part of urban living. On the other hand, many rows between neighbours revolve around the appropriateness of playing loud music at times when it shatters the perceived peace and quiet of a residential neighbourhood. Equally, some community protests (NIMBY or 'Not In My Backyard' campaigns) have been directed towards aircraft noise associated with low-level flying.

An interesting example of how social order is connected to particular understandings of what sounds belong where centres on rural 'noise pollution'. In Chapter 7, we began to explore the way the English countryside has been represented – pictorially, cartographically and discursively – around certain idealized images (such as 'chocolate-box' thatched cottages nestling in a green setting). In turn, the maintenance of this bucolic imagery has relied upon the countryside being protected from certain forces which seemingly threaten a secure, stable and conservative landscape. Matless (1998) has therefore stressed that Modern influences have been allowed into the countryside so long as their 'Modern-ness' is accommodated in the existing landscape order. In the twentieth century, local community responses to electrification, industrial development, windfarms, caravan parks and new house-building in the countryside have therefore been cautiously conservative, generally only supporting development that is in keeping with a picturesque and pastoral aesthetic. More obviously, outright opposition has been expressed by many rural dwellers when 'their' surroundings appear threatened by incursions seemingly originating in the city. In the interwar period especially, while preservationists welcomed the fact that urban dwellers would be able to experience a healthy and morally ordered landscape, they also attacked those working-class citizens whose conduct was unbecoming (Matless, 1995). Litter, flower-picking and disobedient bathing were all frowned upon in this period by those fighting to preserve 'countryside values'.

Yet the threat to the countryside was not just visual, with loudness in the countryside often condemned in the 1930s. For instance, the prominent planner Patrick Abercrombie claimed, 'the honk of the motor car, the sound of the gramophone . . . do not enter into the chord (of the countryside): their dissonance is seriously felt and of singular pervasiveness' (Matless, 1995, 96). The well-mannered citizen, therefore, was one who was noise-conscious, recognizing that certain noises and sounds are not thought to be appropriate in the countryside. This suggests that urban dwellers were supposed to respect the sonic environment which they encountered in the countryside – one of tranquillity. Crang (1998) thus notes that debates about urban noise-makers in the 1930s find their contemporary equivalent in debates about rave culture (Figure 8.7). For example, when young travellers and people from urban areas congregated on Castlemorton Common (Worcestershire) over three days in May 1992, their presence seemed to provoke much anxiety amongst local people as soundsystems and collectives blasted their 24-hour party music into an area defined as an Area of Outstanding Natural Beauty. Protected by a series of local by-laws designed to prevent this landscape being despoiled by mountain-bikers, hang-gliders and dog-walkers, the local authorities were, however, unprepared for the 'invasion' of approximately 20,000 young people determined to turn this public place into one of sociality and entertainment (McKay, 1996). The fact that this was centred on the creation of a social space through the playing of dance music (as opposed, for example, to the music of locally born composer Edward Elgar) seemed to raise further local fears about the presence of this group. Attempts by the police and local authorities to ban the playing of 'repetitive beats music' in public places using the powers of the 1994 Criminal Justice Act represent an attempt to criminalize both rave music and ravers when their presence seemingly threatens social order.

Figure 8.7 Raving: out of place in the countryside?

Photo: K. Wyatt

■ Exercise

Focusing on a park or sports field in a location you know well, try to imagine which groups or individuals might complain if the space was used for an open-air gig or concert. Consider the following:

- Would the time of day or week matter?
- How important would the volume of the music be, or are other factors important?
- Would there be more complaints for certain types of concert featuring different types of music?
- If the local authority know about the concert in advance, would they give permission for it to go ahead? What other considerations would they take into account apart from potential noise pollution?

Ideas about where certain sights, sounds and smells belong are obviously complex, being shaped by a range of power relations between different social groups. Yet often these ideas are taken for granted – for example, we probably wouldn't expect to see a garish fast-food restaurant or advertising hoardings in a National Park, even though they have become an expected part of urban living. Equally, we might not even question planners' decisions that certain styles of housing are acceptable in 'historic' conservation areas (while others are not), or that certain neighbourhoods are suitable for the siting of a refuse dump (but others are not). Yet all of these rely on moral judgements about what is in keeping with the character of a place (and what is not). Tim Cresswell has stressed it is because we take place for granted that we tend not to question who decides what is appropriate where:

> Order is inscribed through and in space and place – the landscape is the truth already established, through the imposition of brick walls, green fields and barbed-wire fences. Through the division of space, 'truth' is established and order maintained. Boundaries and areas carry with them expectations of good and appropriate behaviour . . . to act out of place is to fail to recognise the truth that has been established.
>
> (Cresswell, 1996, 55)

One interpretation of this is that if we fail to recognize (and observe) these powerful, if often implicit, moral codes we are demonstrating our own lack of social competence. Another possible interpretation, however, is that we are deliberately refusing to accept these moral codes as part of an act of resistance, using our power to challenge what is seen as normal and correct. It is this latter interpretation that we will examine in Chapter 9.

■ Reading

Rodaway (1994) offers an interesting account of the role of the senses, which integrates ideas from psychology and sociology to develop a holistic 'geography of the body' (which attempts to combine some of the concepts introduced in this chapter with those discussed in Chapter 3). Cresswell's (1996) 'in place/out of place' explicitly combines ideas about moral geography with those about appropriate behaviour; his case-studies of 'hippies' and female anti-nuclear protesters as misfits certainly provide food for thought. On geography and music, see the special issue of *Transactions, Institute of British Geographers* (1995). Here, a useful overview is provided by Leyshon *et al.*, and more specific topics covered in individual articles by Cohen, Kong, Hudson and Valentine. Additionally, see *The Place of Music* ed. Leyshon *et al.* (1998).

■ Summary

In this chapter, we have begun to explore how the relations between people and place might be shaped by power relations between different social groups and individuals. Key ideas considered here include:

- Power is partly about the ability of the state to control how people ought to act in different places. This control is enforced through policing, surveillance and regulation carried out by agencies and institutions of the state.
- Power is also about self-management and regulation, with people expected to monitor their own behaviour in accordance with ideas about what is desirable and culturally valuable.
- Power is instilled in place through a complex range of laws, norms and expectations (reflected in its design and appearance) which encourage us to adopt particular bodies and behave in particular ways. Rather than being fixed and 'natural', such norms are arbitrary, needing constant maintenance by dominant groups, and varying over space and time.
- Power is not unidirectional, with all people having the potential to resist dominant ideas of what is good and right.

nine Struggles for place

This chapter covers:

9.1 Introduction

In the last chapter we looked at how places are implicated in power relations, with the exercise of power by some groups over others shaping the way in which individuals interact with their surroundings in some obvious (and not so obvious) ways. We saw that the design, organization and use of specific places may be involved in persuading people to act in accordance with certain expectations of what is considered 'good and proper'. Taken together, such a perspective on the relations of place and power might appear to imply that geography is 'carved out' by the rich and the powerful, who occupy places of distinction and prestige while frequently relegating marginalized groups (and behaviours) to areas typified by disinvestment and decay (Sibley, 1995). This assumption is to some extent mirrored in the idea that there is a dominant moral geography of who (and what) fits into certain places (and not others). But the way that this occurs is not as straightforward as it might first appear; as we saw in the last chapter, it is not simply the case that power is wielded in a downwards direction by those in 'authority'. Indeed, while the state and law may act together to lay down norms and expectations of what it means to be a good citizen, these expectations are enacted and maintained as much through self-regulation as through the policing and surveillance carried out by agencies of the state.

One of the things that therefore became apparent in Chapter 8 is that everything we do as human beings (e.g. eat, drink, smoke, walk, sleep, talk – even make love) may be interpreted as caught up in power relations. What we might then begin to think about here is how our everyday actions and practices might challenge or resist dominant meanings and uses of place, deliberately or accidentally changing relations of power. In a sense, this re-emphasizes Foucault's argument that power is everywhere and that it comes from everywhere. As he put it:

> Power is not something that is acquired, seized or shared, something that one holds onto or allows to slip away; power is exercised from innumerable points, in the interplay of non-egalitarian *(unequal)* and immobile relations . . . . Power comes from below; that is, there is no binary and all-encompassing opposition between rulers and ruled at the root of power relations, and – serving as a general matrix – no such duality extending from the top–down and reacting on more and more limited groups to the very depths of the social body.
>
> (Foucault, 1981, 94)

One way we might interpret this complex passage is simply to draw out Foucault's insistence that we all have the power to make and remake the world in particular ways, and that dominant senses of order are constructed through an uneven negotiation between different social groups. In another sense, however, it also stresses that the exercise of power by one group always brings opposition from other groups. As we will see, sometimes these forms of opposition are swept aside, of seemingly little consequence, while in others, they may succeed in changing the nature of the power relations, restructuring the relations between people and place in the process. What all this emphasizes is that places are always struggled over in ways that we do not perhaps routinely recognize or acknowledge. In the rest of this chapter, therefore, we seek to explore how the production, occupation and control of place is caught up in an ongoing struggle between different groups and individuals, taking specific examples of conflict, resistance and struggle to show how these (related) concepts might contribute to our understanding of human geographies. To these ends, the chapter begins by describing the way that the identities of places (like those of people) are riven with tensions and conflicts. Following this examination of how order is established in the midst of complex struggles for place, we turn to explore how particular individuals may set out to challenge this order through political acts of resistance. Finally, we turn to explore the way that places on the margin may be imagined and used as sites of resistance which challenge the values of the centre as we engage with geographical writing on the ambiguous role of 'marginal' places.

9.2 Place, conflict and transgression

In Chapter 5, we looked at the idea that senses of place are not common to all individuals, and that what is regarded as safe and familiar for one person may be another person's source of fear and anxiety. An example we looked at there was the home, often imagined and represented as a haven in a world of change and a place where close, meaningful relationships may be constructed. What this dominant representation misses out of course is the idea that many homes are places where forms of violence, abuse and aggression are common. For instance, some homes are typified by 'dysfunctional' relations between different members of the household, and all too often home is a place where children are on the receiving end of adult aggression or intimidation. This sort of interpretation begins to suggest the existence of a power relation whereby adult values dominate those of children, and where children are 'living in an adult world'. This type of relation is, of course, enshrined in certain laws and expectations about the competence of children at different ages (e.g. in the UK people cannot leave school until they are sixteen, have heterosexual sex until they are sixteen, drive until they are seventeen or drink alcohol until they are eighteen). Collectively, such expectations and norms serve to render children subject to the authority of adults, with parental influence particularly significant for establishing rules as to the sorts of behaviours which are unacceptable in the home at particular times of day. As such, the home may be conceptualized as a space where adult values are dominant and where children's behaviour is limited by their parents' or guardians' concerns for their well-being and 'development' (which may manifest itself in rules about how long they can spend watching TV and playing with their computers, what foods they can eat and what time they should go to bed). These rules may also serve to limit children's use of particular rooms reserved for 'adult' activities at particular times, with the spaces used by their parents or guardians for work or relaxation off-limits on occasions. With children consequently seen as a disorderly presence in these adult spaces, they may find themselves excluded to their own space for long periods of time (see Section 5.2).

Yet this interpretation of adult control is obviously too one-sided, as the home is also a place where children attempt to chisel out their own meaningful (and private) places. This may involve attempts temporarily to transform adult spaces into 'their' spaces through elaborate and imaginative rituals of play such as constructing a 'den' underneath the dining room table or study desk. More prosaically, children may often seek to personalize those spaces designated as belonging to them and make them feel more like their own (by adorning their walls with pictures of sports stars or the Spice Girls, for example). The possibilities for conflict here are multiple, as attempts to mark out a space as their own may be subject to further limitations laid down by parents concerned about the type of images which 'their' children are consuming, let alone the way they are defacing 'their' walls. Similarly, while some children are given their own room which they seek to personalize and fill with meaningful objects, the type of surroundings that a child wants may be seen as chaotic or dirty by adults, who again seek to impose order by periodically tidying or redecorating the room.

But if the home is often experienced by children as an ordered and constraining place, their occupation of space beyond the home is similarly subject to limits. Again, based on assumptions that children are socially incompetent and cannot handle the rigours of navigating public space in an era when the streets are depicted as inherently dangerous, children are largely restricted to occupying officially designated and designed playgrounds (Valentine, 1996). These playgrounds normally contain a collection of single-function play equipment which is essentially safe and predictable (Figure 9.1), and although there is opportunity for children to use them for imaginative play, Matthews (1995) contends that they are essentially sterile environments (see also McKendrick, 1999). Such arguments also extend to include the opportunities for play that exist in formal settings like schools, day-care centres and community settings. However, beyond these 'ghettos' of childhood, it is apparent that children may adapt a range of adult places beyond the home for imaginative and informal play activities. Local wildscapes, wasteland, open fields and woods all seemingly provide children with the opportunities for creative play (and social development), but here again, parental control wields a strong influence. Concerns over safety seemingly inform the way that parents allow their children to venture into different settings unaccompanied, with friends or with siblings, and (as we noted in Chapter 3) there is a strongly gendered dynamic at work here with girls often facing more restrictions. Of course, children may bargain with parents (e.g. claiming that they should be allowed to go to somewhere because

Figure 9.1 A typically sterile and ordered children's playground

Photo: Authors

'all their friends do') or simply disregard their guidance (risking punishment). In either case, what we can discern here is a negotiated geography between adult and child played out through modalities of dependency and autonomy (Matthews, 1995).

As they get older, children and teenagers may begin to contest adult authority more directly simply by hanging out on the 'streets'. Valentine (1996, 213) argues that although the spaces of the street are usually regarded as 'public' rather than 'private', for teenagers they might be the only autonomous spaces they can carve out for themselves (particularly at night when most adults have retired to the sanctuary of their home and may seek to exclude teenagers from 'their' living space). In this sense, the congregation of teenagers on street corners, in bus shelters or in parks may be both an unconscious and a deliberate contestation of adult space, representing teenagers' attempts to define their own ways of socializing and using space away from the spaces (and gazes) of adults. Unsurprisingly, this form of sociality is a source of much concern to adults, especially in an era where teenagers are stereotyped as a threat to social and moral order and associated with drug-use, alcoholism and petty criminality. In view of this perceived threat, teenagers' occupation of space is increasingly curtailed by the police and private security guards who regularly evict 'youths causing annoyance' on the flimsiest pretexts. At the same time, seemingly draconian curfews have been experimented with in many cities as a way of reasserting 'adult authority' (with two-thirds of major USA cities now having some sort of laws concerning children's presence on the streets between dusk and dawn).

An example of how the state, police, security guards and other citizens conspire to reproduce the hegemony (see Chapter 8) of adult space can be seen in the way that skateboarders find their occupation and use of urban space subject to certain limits. Originating as an offshoot of 1960s surf culture, skateboarding offers a distinctive form of sociality to teenagers (and some younger adults) which provides them with a unique way of engaging with adult space (see Figure 9.2). Although some cities provide skate parks and ramps in an effort to confine skateboarding to particular sites, most skateboarders avoid these parks and instead focus on adapting urban plazas, precincts and streetscapes that offer them interesting and challenging opportunities to display their skills to each other (and passers-by). While in some cases skateboarders have little option than to take to the streets (Borden, 1996), it seems that most skaters are deeply aware of how their actions subvert the intended uses of city space, taking over adult space physically and conceptually. This challenge may also be symbolized through the skater's adoption of particular 'streetwear' fashions, language and music styles that seemingly mock established adult values and tastes. As a result, this seemingly anarchic activity has also been subject to forms of surveillance and control designed to remove skating from many city centres, ostensibly to prevent erosion to buildings and street furniture. The demarcation of exclusion zones for skating has, therefore, become a marked feature of many city by-laws:

Figure 9.2 Urban transgression: the skateboarder in the city

Photo: © Simon Shepheard/Impact Photos

> No person upon roller skates, coaster, skateboard, roller blades or similar type device shall be permitted in the central business district, being the area bounded by Illinois Road on the South, Wisconsin Avenue on the North, Oakwood Avenue on the West, Western Avenue on the East, and the length of Western Avenue extending from Westminster Avenue North to Woodland Road, or the western business district being the area bounded by Everett Road on the South, Conway Road on the North, Waukegan Road on the East, and Telegraph Road on the West. For the purpose of this ordinance the central business district and the western district shall include all the properties zoned or used for business and governmental purpose on either side of the streets described above.
>
> (Forest City, Illinois, Ordinance 93–46, 1996)

Of course, the success of these ordinances is very mixed, and for many skaters, such attempts to exclude only strengthen their resolve to 'reclaim the streets', turning the challenge of dodging security guards into something of a game.

This brief description of the negotiations that occur between adults and children/teenagers begins to indicate that while most places can be conceptualized as adult spaces, this production is by no means natural, being struggled over on an almost daily basis. While the struggle is far from even (after all, children are marginalized within the design and planning process, lack political representation and are often financially dependent on adults) it is evident that children have the potential to challenge the parental and adult constraints that confine their occupation and use of space. Equally (but differently), other groups whose lifestyles and interests do not conform to the hegemonic values dominant in society – ethnic minorities, gays and lesbians, the disabled, gypsies, drug-users, prostitutes and so on – may also routinely challenge the way that their use of space is confined by dominant social values. This type of **transgression** (literally, the crossing of geographical and social boundaries) principally occurs when particular phenomena (individuals, groups or happenings) appear as 'out of place', challenging taken-for-granted expectations about what is supposed to be found where (Cresswell, 1996). When such transgressions occur, dominant social groups may act hysterically (through the media, the judiciary and political institutions) creating a **moral panic** about a particular group or practice which acts as a trigger to new forms of social and spatial control. Here, for example, we should note that new attempts to curtail children's and teenagers' access to public space have been fuelled by media accounts of children as dangerous and out of control (see Valentine, 1996).

Of course, transgression can be unintentional, resulting simply from individuals seeking to use particular places as sites of economic reproduction or sociality in a way that inadvertently offends or challenges dominant expectations. Hence, while skateboarders may set out to challenge the meanings and codings of the streets in a manner that deliberately (but fleetingly) flouts conventions, the street homeless may find that their forced occupation of public (and quasi-public) spaces brings them into continual conflict with the forces of law and order. As Daly (1998) points out, the street homeless are a group who are forced to occupy public space for both economic and social reasons (e.g. bartering, begging, socializing, finding a place to live, etc.) because they have been evicted from the private spaces of the real-estate market. However, their subsequent presence on the streets is fiercely contested, and although it is possible to detect a great deal of general sympathy for those living on the streets in Western societies, at a local level this tends to translate into a concern that the homeless should be removed from particular residential and commercial areas. This urge to exclude the homeless has complex origins revolving around people's fears of difference, but certainly in societies where private property is highly valued, those without houses are often spurned as nonentities (Shurmer-Smith and Hannam, 1994). The stigmatization of the homeless thus resonates with the same kinds of metaphors that have been used to describe unwanted and troublesome 'others' throughout history, with ideas of the homeless as dirty, deviant and dangerous provoking fears about the potential of homeless people to 'lower the tone' (and the house prices) in a particular area.

Moreover, in an era when city governors are seeking to attract global investment through vigorous efforts at place promotion (see Chapter 7), the homeless are often being targeted as a group who have no place in city centres catering for affluent consumers, tourists and business-people. By way of example, Neil Smith (1996a) graphically documents the way in which the 'improvement' and **gentrification** of

New York's Lower East Side resulted in the systematic eviction of the homeless from a number of shelters, parks and streets. Most notably, he describes the forcible removal of homeless people from Tompkins Square Park, an area that had been regularly used as a place to sleep by around 50 homeless people, recounting how the police waited until the coldest day of the year to evict the entire homeless population from the park, their belongings being hauled away by a queue of Sanitation Department garbage trucks (reiterating that notions of social purification rest on ideas that certain people are 'polluting'). He sees this heavy-handed treatment as symptomatic of the fears evident amongst the new middle and upper classes who had relocated to the Lower East Side; groups who regarded the unemployed, gays or immigrants as a potential threat to their dream of urban living.

Smith (1996a) describes such heavy-handed attempts to exclude the street homeless from particular sites as symptomatic of the **revanchism** (literally, 'revenge') evident in contemporary cities, where sometimes brutal attacks on 'others' have become cloaked in the language of public morality, neighbourhood security and 'family values'. According to him, manifestations of revanchism have become apparent throughout the urban West with the introduction of curfews, public order acts, by-laws and draconian policing designed to exclude certain 'others' from spaces claimed by the affluent. For example, in many USA cities, this has involved the passing of 'anti-homeless' laws designed to prevent people congregating in parking-lots, making it illegal to panhandle or beg within range of a cash machine or even making sleeping in public an offence (see Mitchell, 1997). Elsewhere, suitable sites for the congregation of the street homeless have been 'designed out' by the authorities barricading public parks at night and removing benches that might be used for sleeping on. Collectively, these acts send out the message that the authorities are concerned to remove homelessness from the gaze of 'respectable' populations, something that is underlined in the espousal of **Zero Tolerance** policing (see below).

■ Exercise

Read the following extract from the *Daily Telegraph* (1997), which describes the apparent diffusion of Zero Tolerance policing from the USA to the UK. As you read, think about who might benefit from Zero Tolerance policing, as well as who might suffer. On this basis, can we say that Zero Tolerance policing is a good/bad thing? Take some time to note down the reasons for your answer, thinking about the power relations that are being played out here.

The drunks and beggars who frequent London's King's Cross found themselves unwitting participants in an experiment on cutting street crime in the weeks before Christmas. The Metropolitan Police put an extra 25 officers on the beat in one of the capital's sleaziest districts as part of Operation Zero Tolerance, a concept given fresh publicity yesterday by Tony Blair, the Labour leader. In his interview with the

Big Issue [a magazine sold by homeless people], Mr Blair became the latest leading politician to lend his support to the ideas pioneered in New York, where they are linked with a dramatic fall in the crime rate. The essential component is that anti-social behaviour will not be allowed. Aggressive begging, drunkenness and intimidating behaviour by 'squeegee merchants' are targeted and offenders arrested. Both Michael Howard, Home Secretary, and his Labour shadow Jack Straw have visited New York to see the system in operation. It was promoted by the city's former police commissioner, William Bratton, who subscribed to the 'broken windows' theory of policing espoused in 1982 by two American academics, James Wilson and George Kelling. It is based on the idea that crime is self-perpetuating: leave one window broken in a building and the rest will soon be smashed. Amid the squalor, a sense of lawlessness will be created that will embolden others to commit more serious crimes. Mr Bratton's approach appears to have paid dramatic dividends. Murders are down 40 per cent since 1993; burglary has dropped by a quarter; there are 30 per cent fewer robberies than two years ago and almost 40 per cent fewer shootings.

London is hardly on a par with New York; but dealing with behaviour that causes fear or anxiety has widespread popular support . . . . A Home Office team reported to Ministers last year that 'anti-social behaviour can stimulate criminality by creating an environment that attracts the more criminally inclined and implies that their conduct will not be subjected to effective controls'. It suggested tackling problems such as sleeping rough, beggars and buskers on the Underground and the anxiety caused to some by squeegee merchants. The Government has recently given police the power to confiscate alcohol from under-age drinkers on the streets and is looking to make begging an arrestable crime if the offender has a fixed address . . . . It is, however, already open to local authorities to introduce by-laws to stop drinking in public – as is the case in Coventry – and to keep their districts clear of graffiti and squalor. Police forces can also use their existing powers to clean up an especially insalubrious area, such as King's Cross – where Mr Blair sometimes drops his children on their way to school and which he said could be 'frightening'. The Metropolitan Police are still assessing the impact of Operation Zero Tolerance but found the high-profile presence of officers on the streets 24 hours a day resulted in a marked improvement in the 'quality of life' for residents and workers. Pete Williams, a spokesman for the police division that includes King's Cross, said that 'the initial reaction was that people were very pleased to see extra officers on the beat . . . there was a dramatic fall in anti-social behaviour in the area'.

Here then, we are beginning to see that there are many conflicts over the use and meaning of different places, so that for some individuals, 'public' spaces are imagined as sites where strangers can come together and celebrate difference, while for others it is important that they are subject to controls that limit their use and occupation. This is as true in the countryside as it is in the city, with the transgressions of New Age Travellers into the common-lands of the British countryside equally prompting attempts to reassert social and spatial order. Inevitably, such attempts to order space involve a complex struggle for one group's 'sense of place' to be imposed on those of other groups, and it is the midst of these struggles that the often taken-for-granted nature of place is revised and reworked (Cresswell, 1996). The type of struggles for place we are describing here have evocatively been termed **turf wars** by Dear and Wolch (1989). This term captures the sense that what is often regarded as 'normal' and proper behaviour in place is struggled over by different groups and individuals who are all seeking to mark off a specific place (or turf) as being available for their type of people to use, excluding 'others' in the process (see also Chapter 5 on territoriality).

■ Reading

The American geographer Don Mitchell has written extensively about the contested spaces of homelessness – see especially Mitchell (1997) – while Neil Smith (1996a) also discusses the way that homeless groups have been removed from particular urban spaces in his wide-ranging commentary on the impacts of economic restructuring and gentrification in Manhattan. The Open University textbook *Unruly Cities: Order/Disorder* (Pile *et al.*, 1999) offers an accessible introduction to debates about the ways in which conflicting senses of place create urban order/disorder, while *Contested Countryside Cultures* (Cloke and Little, 1997) offers a different take on the struggles involved in the maintenance of the 'rural idyll' (see also Chapter 7).

■ 9.3 Place and resistance

Hopefully, the examples of the way that both children and the homeless struggle to appropriate and use different places made you think about the unequal power relations that operate between different people in contemporary societies. Clearly, dominant social values in the West are structured around the lifestyles of home-owning, heterosexual, white, adult males, and both children and the homeless (alongside women, the disabled, homosexuals – and even animals and plants) may find their place in the world subject to a series of controls and constraints that serve to limit their use of space. For example, many homeless people find that their life on the streets becomes a constant game of 'cat and mouse' as they try to 'chisel out' their own places in the face of opposition from local home owners, the police and shopkeepers (as well as other homeless people). Here, the very mobility of the homeless may itself become a form of **resistance** as they seek to challenge attempts at ordering, separating and excluding them from various public spaces. For the street prostitute too, furtive mobility may be important for evading arrest from vice police and harassment from local residents (Hubbard, 1999), while on a different scale the movements of New Age Travellers between urban and rural areas also challenge dominant assumptions that we all aspire to 'settle down'. Accordingly, the work of Deleuze and Guattari (1988) has influenced geographers seeking to theorize the resistive politics of movement and flow, and while their arguments are certainly not straightforward, they begin to highlight the importance of 'nomadic' behaviour in societies where space is often clearly segmented and bounded.

In this way, the idea that geographic order is imposed 'from above' through the panoptic gaze and segregationist strategies pursued by the police, magistrates, engineers and planners needs to be tempered with the observation that at 'street level' we find that individuals and groups create their own geographies, using places in ways very different than bureaucrats and administrators intend. Ideas of resistance are consequently of major importance in many geographers' discussions of power relations, identities and the politics of everyday life. Rejecting notions that power can only be wielded by the powerful, much of this work seeks to document that if the power to own and occupy space is everywhere then it can potentially be resisted everywhere. Many geographers have thus engaged (explicitly or otherwise) with the work of Michel de Certeau, a writer whose attention to the intricacies of resistance has succeeded in highlighting its everyday manifestations and forms. Deflecting attention away from the more 'obvious' forms of resistance – sit-ins, riots, parades, to name just a few – he argued that resistances are more usually assembled from the materials and practices of everyday life. These resistances may be ephemeral and slight; for instance, a single look, a movement or a simple spoken word might all encompass resistance. Developing this logic, de Certeau famously wrote of the simple act of walking as resistive, describing it as a 'process of appropriation of the topographical system on the part of the pedestrian' (de Certeau, 1984, 97). Here, he implied that pedestrians can reclaim the streets through improvisational tactics, with their footfall fleetingly appropriating spaces that have often been rendered 'sterile' by engineers, planners and architects. The act of walking therefore becomes, in effect, one of the principal ways in which citizens can refute the notions of moral and social order which have been inscribed on the landscape. Notable examples here include the way that many women's groups have been involved in attempts to 'Reclaim the Night' simply by walking on the streets of cities at night (Watson, 1999), challenging the assumption that women are not capable of negotiating public space without the assistance and 'protection' of men (an assumption that often restricts them to private space, as we discussed in Chapter 5). Elsewhere, many ramblers have sought to exercise their 'Right to Roam' by trespassing on to private land (see Figure 9.3). In this sense, walking itself is a distinctly politicized – yet obviously routinized – practice whereby walkers are often able to weave their own routes through space in a way that escapes measures of control (and challenges existing orders in the process).

Following de Certeau's arguments, it appears that the most important factor underpinning our understanding of what constitutes resistance is its underlying **intentionality**, something that can only be understood by appreciating the spatial context in which it occurs. By this, we mean that the intention of the actor, in terms

Figure 9.3 Exercising the 'Right to Roam'? The mass trespass on Kinder Scout, the English Peak District, 1932

Photo: Ramblers' Association

of his/her refusal to recognize dominant socio-spatial orders, distinguishes an act of resistance from one of transgression (which may simply result from someone being ignorant as to what actions are appropriate in a given place). As Tim Cresswell has discussed at length, if space and place are used as a means of imposing social order, then intentionally acting 'out of place' is to refuse to play by the rules. As he argues, the 'unintended consequence of making space a means of control is to simultaneously make it a site of meaningful resistance' (Cresswell, 1996, 163). This notion of resistance celebrates the seemingly endless ability of often marginalized individuals and groups to make the best of their situation, mocking and subverting the everyday exercise of power to their own ends. De Certeau (1984) referred to these resistive appropriations of everyday spaces as **tactical**, seeking to distinguish between the strategies of the strong and the tactics of the weak. By this, he means that while the powerful in any given context may classify, survey and order space, the weak can only divert, manipulate and subvert the meaning of specific places. In this sense, he implies that a consideration of practices of resistance cannot be separated from a consideration of practices of domination (such as exclusion or segregation).

Places, then, do not just exist as classified, ordered and labelled as belonging to certain groups by the state and its agencies (the police, planners, councils), but as places whose meaning may be challenged and subverted through everyday spatial practices and tactics. As Rob Shields (1997) stresses, it is important that, in their anxiety to create grand theories of socio-spatial life, geographers do not unthinkingly ignore the small resistances enacted by human beings in the field of everyday life. Even small acts need to be examined in terms of their potential for challenging and redefining dominant geographies. An example here might be the graffiti and tags left by those wishing to claim part of urban space as their own – a defacement which destroys the official meaning of the carefully designed and controlled spaces of the city:

> The tagger, the specialist who leaves his mark on the wall, is a hit and run calligrapher – probably young, MTV-grazing and male. His art is nomadic, a matter of quantity not quality. As often than not, the deed is carried out on the way back from a club in the early hours of the morning; the announcement of a jagged progress across some territory. Nothing too bulky to carry, a good black felt-tip in the back pocket of your Pucca jeans will do the trick.
>
> (Sinclair, 1997, 2)

As Sinclair contends, urban graffiti is often an anonymous autograph but the use of a distinctive tag is significant, parodying the messages and tags plastered across other buildings by big business (e.g. the hoarding advertising Coca Cola, the poster for Levi's jeans). As Cresswell (1992) points out, *where* the graffiti is becomes crucial in defining it as an act of resistance. In an art gallery (or even a youth club), graffiti might be praised as creative, innovative and even artistic; but on the side of a train or on the wall of a building, it becomes a sign of disorder and an act of resistance (see also Chapter 4). The same is of course true of any resistive act, as while a group of young people dancing at a festival like Glastonbury or Lollapalooza is expected and even encouraged, the same sort of gathering on the streets of London's financial district would prompt attempts to reassert order (see below). Appreciating the taken-for-granted qualities of the places where acts of resistance occur is thus crucial in understanding the political consequences of such acts.

■ Exercise

Look at Figure 9.4, which shows a group of protesters confronting riot police as part of a protest against capitalism in the summer of 1999. As you look at this photograph, you might like to consider whether the protesters' actions are justified or whether such political points could be made equally effectively through 'legitimate' channels of protest (i.e. pressure groups, consumer boycotts, petitions, etc.). Is this anti-social behaviour or a valid attempt by a group to point out the inequities of contemporary financial and business practices? It might help to know that this confrontation occurred outside the International Futures and Options Exchange in the City of London.

Figure 9.4 The Carnival against Capitalism (Summer 1999)

Photo: Roy Riley © Times Newspapers Limited

By now, it should be evident that resistance can take a variety of physical, mental and social forms, and that while resistance may be effective at challenging the dominant organization of space when it is active and open, latent and symbolic forms of resistance are also important. Accordingly, many geographers are interested in the resistive potential of bodily performances, that is to say, those presentations of self which are public and embodied. In Chapter 8, we began to show that the presentation of the body projects an image of the person to both the individual themselves and those who observe them. In some instances, those who fail to match up to ideals of what good, disciplined, healthy citizens are supposed to look like seek to develop performances of the body which enable them to cope with the sense of stigma which they experience in public places. The importance of such **coping strategies** was first elucidated by the psychologist Ervin Goffman, who felt that the pressure of idealized conduct could be most clearly seen amongst marginalized populations whose deviance forces them into 'discredited' or 'discreditable' groups, based on the nature of their stigma (Goffman, 1963, 42). For the discreditable individual, attempts to 'pass' as normal in public space typically involve the use of 'dis-identifiers' to establish themselves as 'normal'; this might include pregnant women hiding their pregnancy, disabled people using prosthetics, or women power-dressing to communicate that they belong in the male-dominated worlds of finance and business (McDowell, 1999).

While coping strategies may help 'others' blend in to specific places, feelings of ambivalence and alienation may be experienced by those who feel they are having to present their bodies in ways that are not 'true to themselves'. As such, Goffman also describes how bodies may be presented in ways that challenge taken-for-granted expectations of how we are supposed to manage our appearance in different places. Such explorations of how performances of the self serve to construct or contest expectations about (placed) identity have been most fully described in the work of Judith Butler, who contends that these performances need to be understood in relation to scripts already written through the conventions of society. Her own work has focused on the performance of sex and gender, describing the way we make our bodies intelligible as male/female, gay/straight, through the adoption of certain forms of clothing, make-up, hairstyles and body adornment which serve to 'sexualize' our bodies. This extends to include the idea that the body itself should be sculpted and managed (through diet, body-building, plastic surgery, etc.) in such a way as to conform to dominant expectations of what sexually alluring men and women should look like. Here again, we might think about the way that we 'work' on our bodies, dieting, working out, picking, pruning, squeezing and decorating them to conform to some idealized view of an appropriate masculinity or femininity (Butler, 1990).

In one sense, these ideas suggest that our bodily identities are constructed by the powerful social forces of normalization which we began to examine in Chapter 8. At the same time, however, Butler's work has been extremely influential in alerting researchers to the way that bodily performances may subvert expected norms of public behaviour and appearance. As she describes, often the labour which people invest in their bodies, and the way that they make their bodies perform in specific spaces may serve to oppose dominant expectations by refusing to conform to some dominant standard (for example, by doing unhealthy things to our bodies such as drinking, smoking, over-eating, etc.). Here, we need to think about the body not simply as a site of knowledge-power but also as a site of resistance: a fleshy corporeality that may exert a stubborn recalcitrance and which always provides the possibility of a counter-strategic inscription (Grosz, 1994). Butler shows this by describing how the parody of cultural conventions through **excessive performances** such as that of the ultra-masculine 'gay skinhead', the male drag artist or the feminine 'lipstick lesbian' offers a challenge to dominant heterosexual norms. The sight of two apparently 'macho' men kissing or holding hands in certain places thus disturbs assumptions about what is normal (and where), exposing what Butler describes as the 'fiction' of compulsory heterosexuality. Again, context is all-important here, with these parodies having different effects dependent on the type of spaces where they occur (McDowell, 1999).

While Butler's ideas of performativity have largely been developed in the context of performances of sex and gender, they are also relevant in elucidating the importance of performance, dress and style in other situations. For instance, there is now an extensive literature on **subcultures** which has considered the significance of rituals of resistance as symbolic (rather than practical) forms of opposition. In general terms, a subculture is usually conceptualized as any faction of society whose practices and lifestyles are defined in opposition to dominant ways of life, and whose adoption of particular fashions, music styles and modes of communication marks them off as different from the mainstream. While people of different ages may belong to such subcultural groups (and we can perhaps think here about the way in which some people devote their lives to religious or political cults) most research on subcultures has focused on the way that the young use consumption and style to unite in oppositional cultures and communities. Influenced by the pioneering work of Stuart Hall and the Centre for Contemporary Cultural Studies established at Birmingham University in the 1970s, many writers have focused on the musical tastes and dress codes of diverse and often short-lived youth movements (e.g. mods, rockers, parkas, punks, rastas, new romantics, hippies, etc.). Theoretically, this work draws attention to the way that consumption and style are used by the young to represent their opposition to and rejection of adult values. An example here would be the way that punks used safety pins as earrings or body-piercings. Here, an object ordinarily used to fasten a baby's nappy was taken out of context and transformed into a fashion item, negating normal expectations about the value and meaning of this commodity. In the context of punk, the appropriation of this object sent out a signal that here was a movement that refused to accept dominant meanings, that questioned the way things were and sought to develop new possibilities for social life (Hebdidge, 1979). At this point, you might begin to think about how more recent fashions – skatecore, goth, grunge, bhangra, hip-hop, rave, ragga, big beat, gangsta rap and so on – have been associated with (street)styles which subvert or mock expectations about what people should aspire to look like (privileging 'authenticity' over the trends promoted by the fashion industry and/or adopting 'mainstream' fashion to make their own style) (Figure 9.5).

Yet more recent work on subcultures has sought to problematize the idea that individuals can be said to belong to one 'unified' subculture, stressing that young people rarely adhere to just one style of music or one way of dressing. Recognizing that young people may adopt 'mix-and-match' lifestyles, taking elements from different fashion scenes and movements to create their own style and identity, it becomes difficult to accept that all people imbue a particular record or style of training shoe with the same cultural meanings. As such, it is perhaps dangerous to say that a particular ritual of youth identity and belonging is always conceived as an act of resistance, as the meaning of youth styles may differ massively from place to place and from person to person (e.g. buying a record by the Wu-Tang Clan may symbolize entirely different things to an African-American living in Compton or Queens than to a white Protestant American in Oklahoma or a Japanese student living in London). Engaging with this fragmentation of youth identity, Maffesoli (1989, 9) has written of the proliferation of **neo-tribal** identities in contemporary society, noting that the 'polydimensionality of lived experience' has resulted in the emergence of more diffuse and hybrid social groupings as people search for 'meaningful belongingness'. This theorization stresses that people can (and do) move between different 'tribes' rather than having their identities determined by their membership

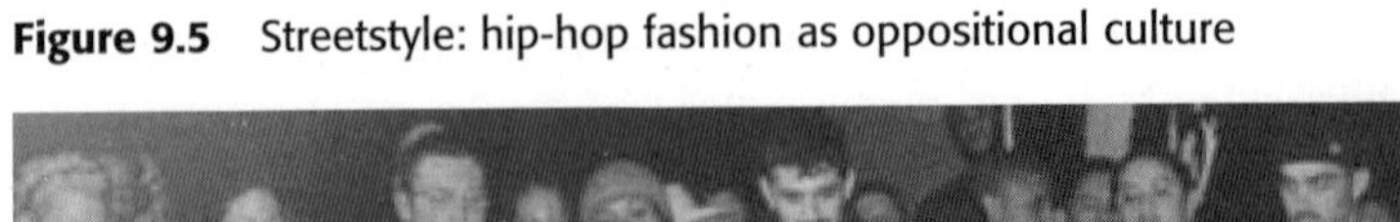

Figure 9.5 Streetstyle: hip-hop fashion as oppositional culture

of a single subculture. These neo-tribes, then, are inherently 'messy', with their rituals of resistance, styles of dressing and favoured musics being much more varied than was proposed by the Birmingham school (see Baldwin *et al.*, 1999). Yet, as with subcultures, opposition and resistance may still be an important focus for neo-tribal activity, even if the type of pleasure which the members get from this activity varies massively. While the 'DIY' politics of, for example, ecoprotest, pursued by many contemporary neo-tribes remain subterranean, hidden in a litany of Web sites and fanzines, these politics may rise to the surface when these 'tribes' come together in spaces that serve to mock the mainstream. It is the resistive potential of marginal places that we seek to explore in the final section of this chapter.

■ Reading

McKay (1996) offers an expansive overview of the types of counter-culture which have emerged since the 1960s, showing how these groups have challenged dominant interpretations of space through a mixture of love-ins, rave-ups and spatial transgressions. The collection edited by Gelder and Thornton (1997) offers perhaps the best overview of the varied debates on the formation and meaning of subcultures, while the spatialities of resistance are critically explored in Pile and Keith (1997).

9.4 Speaking from the margins: the cultural politics of place

As geographers have begun to unravel the complexity of power relations being played out in Western societies, it has become more and more apparent that attempts to impose order 'from above', socially and spatially excluding undesirables from valued and privileged places, are resisted in the routine politics of everyday life. Often inadvertently, marginal groups may challenge dominant expectations by transgressing into places where they are not expected, where they do not 'belong'. As such, the distinction between the centre and the margins is less clear-cut than administrators and planners might imagine, being cross-cut by a more fractured set of divisions which problematize the coding of space as central/peripheral (see Chapter 6). David Sibley (1998) has therefore contended that power relations cannot merely be inferred from the 'facts' of location; the powerful are not always those who seek to control the centre, while those on the margins are never powerless. In some cases, people might even prefer to live on the margins – their exclusion is thus self-imposed.

Extending this logic, a number of geographers have begun to think about the potential of places on the margin to act as sites from which the relatively powerless can organize themselves into self-supporting cultures of resistance and co-operation. Much of this work is influenced by the radical writings of **post-feminists**. This term groups together a large number of writers who do not wish to see women treated the same as men (as is the case in much research conducted by feminists), but instead want to celebrate gender difference, allowing women a speaking position on their own terms, unconstrained by the strictures of male-dominated academic convention. Authors like bell hooks (whose pen-name alone indicates a rejection of academic conventions) and Gayatri Spivak have thus been influential in encouraging geographers to question the 'real and imagined' role of place as a ground for political struggle, alerting them to the complex cultural politics being played out on the 'strategic margins' (e.g. Soja, 1996). Moreover, other 'other' voices have emerged from the margins of geographical enquiry as people of colour, gays and lesbians, the disabled and religious minorities fight to have their geographies rewritten in ways that do not impose order, constancy and predictability on their lives through geographers' desire to identify and theorize differences between 'same' and 'other' (Smith, 1999).

What we are beginning to see as a result is the emergence of geographies which affirm and make space for 'other' discourses in ways that respect details and difference, fragmentation and chaos, substance and heterogeneity in a manner that is almost entirely antithetical to the formal geometries of spatial science (Philo, 1992, 159). The contours of this critical **post-structural** geography are still being discussed, debated and contested by geographers divided as to whether this will be excitingly productive in terms of new ways of thinking about space and place, or will lead to a morally bankrupt geography that is unable to say anything meaningful about anything (e.g. contrast Dear, 1988; Philo, 1992; Warf, 1993; Murdoch and

Pratt, 1993; and Peet, 1998). Although these debates are interesting (and we will return to them in Chapter 10), here we simply want to show how this concern with difference has heightened awareness of the multiple forms of resistance which serve to rupture and **deconstruct** easy understandings of centres/margins. One good example of this is in recent geographic research on the way that sexual identities are constructed and contested. There is now an assortment of critical geographic scholarship which has indicated that places are fundamentally shaped by the dynamics of human sexuality, reflecting the ways in which sex is represented, perceived and understood (Knopp, 1995, 149). These arguments have been most forcibly made in research by geographers on the experiences of lesbians, gays and bisexuals which have collectively suggested that many places are experienced by such groups as aggressively heterosexual (Adler and Brenner, 1992; Bell, 1992; Valentine, 1993). Hence, while displays of heterosexual affection, friendship and desire are regarded as acceptable or 'normal' in most places, it has been noted that homosexuals are often forced to deny or disguise their sexual orientation except in specific (and often marginal) places because of fears of homophobic abuse and intolerance. Yet simultaneously, it has been recognized that the appropriation and transgression of heterosexual spaces may be a potent means for lesbians, gays and bisexuals to destabilize and undermine processes of homophobic oppression as they adopt a variety of tactics in order to challenge the dominant production of space as 'straight'. The establishment of 'queer spaces', frequently (and perhaps misleadingly) referred to as gay or lesbian 'ghettos', has been recognized to play a fundamental role in this process, with such sites often acting as bases for the social, economic and cultural reproduction of gay male and lesbian communities. Areas in the USA (such as Castro in San Francisco, West Hollywood in Los Angeles and Provincetown on Cape Cod), in particular, have been transformed from marginal spaces of persecution to relatively affluent centres of gay cultural life through political organization, creativity and activism. In these areas alternative economies may thrive (particularly in the cultural industries – food, music, fashion, arts and media), with gay entrepreneurs bringing different values and ideas to the attention of wider audiences, making them more mainstream in the process. This gradual **demarginalization** process may, over time, bring such marginal places closer in character to those in the centre, although many commentators remain cynical about the basis on which gay lifestyles are made into commodities to be bought and sold (or whether this movement is desirable anyway).

Yet perhaps the most important notion underpinning writing on geographies of homosexuality has been that interpretations and understandings of sexual subjectivity should be based on 'close' work rather than that conducted at 'arm's length' (Bell, 1992, 328). This involves geographers developing a critical awareness of the **situatedness** of knowledge, aiming (where possible) to replace the traditional academic ideal of presenting an objective 'view from nowhere' (see Haraway, 1991) with a queer objectivity that embraces the heterogeneity and multiplicity of gays' and lesbians' lives. Part of this process has involved geographers making connections between the production of non-heterosexist knowledge and the practices and politics of doing research. Methodologically, this involves the use of research methods designed to challenge 'conventional' research techniques, allowing lesbian and gay research subjects to tell their own stories (i.e. their situated 'views from somewhere') rather than having heterosexist assumptions foisted upon them (Table 9.1). As such, the literature on geographies of sexuality is often highly politicized, with gay and lesbian geographers seeking to destabilize and critique geographical knowledges which oppress them personally. In this respect, the influence of **queer theory** has also been of major importance. Although highly contested, queer theory (or, more correctly, theories) involves a radical attempt to invert simple expectations of gay and lesbian identity by making the mainstream queer through strategies of parody, subversion, camp, androgyny and homoeroticism. Queer theory demands that the 'dispassionate' academic voice is replaced by a variety of other voices in accounts of sexuality as the geographer becomes a political activist; queer geographical theory not only explores the (sexed) relationships between people and place, it provides possibilities for new lesbian and gay identities. From a queer perspective, the real and imagined geographies of everyday life

Table 9.1 A comparison of conventional/heterosexist and alternative/feminist research methods

	'Conventional'	'Alternative'
Topic of study	Issues derived from scholarly literature.	Socially significant issues arising from research.
Data type	Reports of attitudes and actions in questionnaires and surveys.	Feelings, behaviours, thoughts, insights and actions as witnessed or experienced.
Units of analysis	Predefined concepts.	'Natural' events/ communities.
Researcher's relationship with research setting	Attempt to manage and control research role.	Openess, immersion in research situation.
Researcher's relationship with research subjects	Detached.	Involved, participatory.
Implementation of methods	Decided in advance.	Method determined by characteristics of research setting.
Validity of research	Evidence, statistical significance, replicable research.	Understanding, illustrativeness, plausibility.

Source: Adapted from McDowell (1999)

become intriguing spaces for the celebration of 'perverse' sexualities; the relationship between sexuality and place is malleable, offering moments for queer transgression. Peet (1998) accordingly asserts that queer theory involves the development of a radical approach to geography, a way of knowing and (de)constructing people–place relations based on a queer reading of 'mainstream' geography. In many respects then, a queer geography represents a rejection of heterosexual ways of ordering the world and involves an active critique of how places have been studied in 'traditional' geography (in a similar manner to the way that post-feminist theorists have insisted on attacking the masculinist basis of geography and **post-colonial** theorists have attacked its whiteness. Other situated accounts might arise from varying positions of, for example, age, social class or (dis)ability).

■ Reading

On the geographies of sexuality, the key reference is the varied and stimulating collection edited by Bell and Valentine (1995b). A shorter article examining the rapid growth of work in this area is provided by Binnie and Valentine (1999). One of us has also sought to extend this analysis of the connections of sex and place by questioning how everyday places shape our sexual identities (and bodies) whether we are straight or gay (Hubbard, 2000); you might want to look at this if you feel that 'queer theory' is only of interest to 'queer' geographers.

A slightly different take on the forms of politics played out in the margins is offered in Foucault's suggestion that certain places might be conceptualized as **heterotopias**. Although frequently invoked by geographers intent on documenting the resistive potential of particular places, this concept is one that remains vaguely defined and, accordingly, has been carelessly applied to a range of different places. It has been contended that this is largely because Foucault's all-too-brief introduction tended to suggest a range of exciting possibilities but was unspecific in terms of what these conceptual possibilities mean 'on the ground' (Philo, 1997). Yet the vagueness implicit in this concept does emphasize the ambiguous role that these sites play in social ordering. To quote Foucault:

> Heterotopias . . . have a function in relation to all the space that remains. This function unfolds between two extreme poles. Either their role is to create a space of illusion that exposes every real space . . . as still more illusory. Or else, on the contrary, their role is to create a space that is other, another real space as perfect, as meticulous, as well arranged as ours is messy, ill-constructed and jumbled.
>
> (Foucault, 1986, 27).

Rather than straining to identify Foucault's intention to point out the dual role of heterotopias, however, most commentators utilize the concept to highlight the properties of heterotopias that challenge dominant modes of spatial ordering. Specifically, in heterotopias it is the juxtaposition of objects, activities and people not normally found together that is suggested to challenge hegemonic ways of regulating and representing space. This convergence of such miscellaneous and discordant things, it is argued, may erode people's sense of security, disrupting common-sense meanings of space. As such, Hetherington (1997, 8) writes that there is nothing implicit in a site that leads it to be designated as heterotopic; rather, it is what it represents in contrast to other sites. Hetherington uses the example of how gatherings of travellers and druids at Stonehenge transform this site into a heterotopia as pagan rituals and worship of the earth goddess bring a number of disparate (neo-)tribes together into a **'carnivalesque'** festival that mocks conventional Western values. Another example that Chris Philo has used to illustrate what is meant by a heterotopia is the appropriation of land by religious communities such as those distinctive communities established by the Mormons, Plymouth Brethren and Amish people. Such communities, in different ways, are seen to embrace values which conflict with mainstream understandings of right and proper ways of living. For example, the Plymouth Brethren reject ideas that progress should be equated with technological advance, and seek to create insular communities uncontaminated by contact with television, computers, newspapers and other media which, they argue, are used to spread immorality. Similarly, Bell and Valentine (1995a) document how lesbian-feminist groups have used communal living as an alternative which challenges dominant patriarchal lifestyles. As such, these places can be conceptualized not so much as sites of resistance but as sites that might provide the model for alternative modes of social and sexual ordering – as Hetherington (1997) puts it, heterotopias are seditious and mysterious places where new social orders can be 'tried out'.

This idea of heterotopia reminds us that marginalized groups continually (and inevitably) 'speak back' to dominant modes of power by being located in (and locating *themselves* in) spaces which mock mainstream society's attempts at creating and policing boundaries, categories and partitions. A somewhat similar idea is invoked in Hakim Bey's (1991) theorization of a **Territorial Autonomous Zone** (TAZ), a term which he uses to refer to those locations which are appropriated (in an often ephemeral manner) and transformed into experimental, anarchistic places where new forms of governance can be tried out. This concept of a TAZ is based on Bey's trawling of history to identify a number of distinctive and subversive communities which exploded briefly and brilliantly only to disappear just as quickly. For example, Bey writes of how the sea-rovers, pirates and corsairs of the eighteenth century created an 'information network' that spanned the globe. While primitive and devoted primarily to a rather grim business, the net nevertheless functioned admirably for their purposes. Moreover, scattered throughout the net were islands, remote hideouts where ships could be watered and provisioned, booty traded for necessities and luxuries. Some of these islands supported 'intentional communities', whole mini-societies living consciously outside the law and determined to keep it up, even if only for a short but merry life. For Bey, these were archetypal TAZs, guerrilla operations that did not engage intentionally with the state, but liberated a space only to later relinquish it. Ingham *et al.* (1999) claim that the contemporary equivalent of the TAZ is the warehouse rave, an unlicensed event where a combination of drugs (ecstasy and speed) and repetitive-beat (trance) music alters accepted senses of space and time, bringing participants together in a place which transcends contemporary social norms.

■ Exercise

While it is possible to chart the emergence (and disappearance) of events, communes, gatherings and happenings which could be described as Territorial Autonomous Zones, it should be noted that discrepant behaviours and identities are also played out in cyberspace. Surfing the Internet, you may come across many Web sites, multi-user domains and discussion fora which bring together ideas, images, words and sounds in ways that challenge conventional understandings. Try using search engines to look for sites relating to gay rights, anti-road building or green consumerism. To what extent do any of these sites share characteristics with Bey's concept of a TAZ? Do they contain any material that might be condemned by mainstream society? Are they permanent or temporary? Are these 'spaces' that exist outside of social and legal norms? Ultimately, do they challenge the mainstream or not? You might want to look at Bassett (1997) and Chapter 4 of Kitchin (1997) for some clues on this.

Thinking about the power relations being played out in cyberspace, we are inevitably faced with the problem of whether we can reduce resistance to a single effect. Certainly, although many of the Web sites you may have looked at challenge the orthodoxies of language (as well as the law), you may begin to think that the Internet is something of a legitimate site of protest, a 'democratic' place where subversive, seditious (and perhaps pornographic) material is expected to be encountered. This again serves to remind us of the importance of context, as some transgressions and inversions of everyday life clearly do not disturb existing power relations. Indeed, some forms of carnivalesque behaviour are even licensed and encouraged (and at the same time policed and rendered 'safe') by the state. On this basis, Stallybrass and White (1986) have contended that it makes little sense to talk about the inherent radicalism or conservatism of acts of resistance. Instead, it is necessary to acknowledge that acts of resistance may be conservative in one place and radical in another, or even contain elements of both simultaneously. Here then, we need to examine carefully the way that power relations are constructed in the midst of a complex cultural politics in which the meanings and identities of place are fundamental.

These observations about the cultural politics of resistance serve to highlight the way that many geographers are now engaging with the complexity of the world rather than forcing it to fit into established geographic models/theories. Conflicts over the use of place, and the associated creation and destruction of different senses of place, are therefore something that geographers are beginning to embrace rather than wish away. In this sense, many commentators have praised geographers for their enthusiasm in incorporating 'other' world views and 'other' voices into their writings (e.g. Duncan, 1993; Gregory, 1994; Sibley, 1995). While issues of identity and representation are clearly slippery ones (see Chapter 7), it could be argued that much of the 'cultural' geography cited here gets to grips with 'otherness' in a way that does not relegate the lives and aspirations of 'others' to the footnotes of geographical endeavour. The stress here is on the multiplicity and indeterminacy of place, for, as we have seen, place is experienced, represented and conceived in conflicting ways by different individuals and groups. This final point is a vitally important one, because if geographers de-couple their investigation of human geography from the specificities of the people–place relations being played out in particular contexts, they risk constructing geographical knowledges that place them in an unethical relationship with those beyond academia. It is this debate about the ethics and politics of geographic research that we return to in the concluding chapter.

Reading

Jackson (1989) still represents one of the most accessible and convincing introductions to the way that the meaning of place is constructed and contested in the realms of cultural politics. Post-structuralist approaches to geography – which flatly reject notions of truth in favour of a deconstruction of truth claims – are certainly difficult to get to grips with, though Shurmer-Smith and Hannam (1994) and Doel (1999) have undoubtedly been important for disseminating these ideas to a wider audience (in very different ways).

Summary

In this chapter, we have begun to look at the ways that experiences, representations and perceptions of places are the outcome of a complex and often 'lopsided' struggle between different social and cultural groups. While these struggles are complex, we have tried to stress a number of key arguments here, including:

- Power is not something that is held by an elite, who carve up the world as they would wish; rather it is exercised from innumerable points and involves a constant struggle between different social cultural groups.
- Conflicts over the meaning and use of place are of major significance in reaffirming or contesting existing power relations; as such, transgressions may prompt attempts to reassert social and spatial order.
- Transgressions may be obvious, physical disorderings of place or subtle, symbolic parodies of convention. The impact of either, however, depends on who is transgressing and who notices (both of which are place-specific).

- Generalizing about the meaning of resistance is clearly problematic; struggles for place can therefore only be understood in relation to the specific cultural politics of place.

ten Departures . . .

This chapter covers:

10.1 Introduction

At the outset it was stated that this book aimed to explore different ways that geographers have sought to examine the everyday interactions that occur between people and place. Of necessity, we have travelled quite quickly through a variety of different geographical literatures, alighting at various points to see how different writers have theorized, studied and generally made sense of the relationships that exist between people and place. In essence, each chapter hints at a different way that geographers have sought to conceptualize (or *imagine*) this relationship. While it must be appreciated that each perspective or approach is a great deal more complex than can be described in one chapter, we hope that we have at least given a flavour of some of the more common ways that geographers have sought to make sense of the world. Starting by questioning the reductionist impulses evident in spatial science, the various chapters in this book have examined a number of ways that human geography has been re-peopled, focusing on the following ideas or theories:

- that interactions between people and place are mediated through the cognitive processing of environmental information (Chapter 3);
- that place is constructed subjectively, with different senses of place emerging from the individual relations that people have with their surroundings (Chapter 4);
- that senses of place differ between members of different social groups, so that one person's place of safety and refuge is another's place of fear and danger (Chapter 5);

- that understandings of the meanings of place emerge from widely shared (and socially produced) myths based on understandings of 'self' and 'other' (Chapter 6);
- that place myths are created by cultural agents who seek to represent place in texts, media images and other cultural artefacts (Chapter 7);
- that places are bound into complex power networks that encourage people to behave in certain ways in particular places, disciplining bodies in the process (Chapter 8);
- that experiences, perceptions and representations of place are struggled over and contested between different groups in the cultural politics of everyday life (Chapter 9).

At the end of this journey, therefore, it might be reasonable to expect some sort of conclusion, an ending that makes it clear what we have learned along the way, explaining how these different ideas combine to provide the basis of a coherent contemporary human geography or proposing that one idea is inherently 'better' than another for making sense of the world.

Unfortunately, it is extremely difficult, if not impossible, to suggest that there is one approach (or even combination of approaches) that best enables human geographers to make sense of the relationship between people and place. If anything, what we have learned here is that nothing in human geography can be taken for granted, with uniform, 'common-sense' notions of what we mean by even basic terms like 'people' and 'place' actually dissolving in the face of vigorous debates about the type of relationships that exist between society and space, social structure and human agency or nature and culture. Take the notion of the person, for example. Here, it might seem obvious what we mean when we talk about 'people' – after all, we are all (presumably) human beings whose status as people seems straightforward and unobjectionable. But, as we have proceeded through the different ways that human geographers have approached their studies, it has perhaps became apparent that people have been conceptualized in very different ways as geographers grapple with questions about what makes an individual truly human. Different models of 'man' (and, latterly, woman) have consequently been developed by geographers anxious to put the 'human' back into human geography. Although these have largely succeeded in ousting the 'peopleless' models of positivist spatial science (see Chapter 1), they have left us with a variety of different conceptualizations of humanity. As we have seen, people have been imagined variously as knowing subjects, decision-making machines, irrational actors, embodied subjects, a collection of memories, members of social groups, cultural agents and so on. As you have read through the chapters you have perhaps thought about these different ways of thinking about humanity and considered which best captures what it means to be a human *being*. This raises a difficult set of questions about the essence of being. For example, are you defined by your mind, with its capacity to reason, to imagine and to dream? Or are you constituted by your body, whose fleshy physicality defines your presence? Is it your social identity that defines you, in the sense that you are deemed to belong to particular gender, ethnic and class categories? Do you act rationally or do unconscious desires impel you to act in ways even you cannot understand?

Ultimately, this represents an irresolvable series of questions, as definitions of selfhood are ever-changing and contested in relation to constantly evolving suppositions about bodies, identities and subjectivities (see Pile and Thrift, 1995; Cloke, 1999). For instance, many geographers are currently questioning whether it is even possible to talk about the agency of human beings as distinct from that of other animals, objects and forces in an era when some are proclaiming the 'death of the (human) subject':

> The very speed of the new technological order destroys the horizon, devastates being, paralyses subjectivity in a death-like stasis to be bombarded by signal-objects in electromagnetic fields. This is the scenario ever repeated in turn of the century popular culture where there is no longer a constitutive outside of a swirling vortex of microbes, genes, desire, death, onco-mice, semiconductors, holograms, semen, digitised images, electronic money, and hyperspaces in a general economy of difference. This is a scenario in which even being is reduced to the status of an actant, of equal status with the other main types of actant with subjects, objects and symbols.
>
> (Lash, 1999, 344)

Here, simple understandings of human beings as autonomous knowing agents (i.e. entities who are profoundly aware of themselves and their power to shape the world) break down because of the difficulty of distinguishing between humans and cyborgs (or between nature and the genetically modified). In a world where some human traits – touch, sight, taste, memory – can now be simulated (Haraway, 1991), it is perhaps more necessary than ever to think about how we incorporate understandings of humanity into the geographies that we write (Thrift, 1994). As such, some geographers are looking towards **actor-network theory**, a perspective that suggests that the world is shaped by 'actants' – human and non-human entities – which hold each other in positions that define their identities and ability to *do* things (see Murdoch, 1997; 1998).

Questions about how geographers might best define 'people' are obviously complex, being informed by geography's constant dialogue with the other social sciences (e.g. psychology, sociology, cultural studies, women's studies, etc.). Equally (but differently), we might also pause here to think about the definition of 'place'. At the outset, we began with a fairly unproblematic definition of place as a portion of the Earth's surface – a location in space marked off by boundaries imposed by people (walls, roads, political boundaries and so on). This view of place imagines it primarily as an objective phenomenon, a physical 'given'. While this represents one way of looking at place, it has become evident that other views are also possible (and perhaps necessary). Indeed, Ed Soja (1996, 71) argues that this objective, 'empirico-physical' view of place downplays the importance of place in the formation and reproduction of human relations, portraying it merely as a surface upon which human behaviours and social actions are played out. In the preceding chapters then, we have begun to examine the development of different conceptualizations of place, suggesting, for instance, that they are subjective spaces bought into being through human consciousness; that they are settings produced through the intersection of social practices; that they are repositories of human emotion; or that they are cultural products bought into existence through the contested realms of representation. Here then, we can discern conflicting interpretations of place – as real or imagined, objective or subjective, physical or mythical.

Despite the seeming incompatibility of these different interpretations, much human geography over the last three decades has been preoccupied with breaking down and reworking these complex conceptualizations of place. This has by no means been a straightforward task, with Nicholas Entrikin (1991, 12) having argued that 'to understand place in a manner that captures its sense of totality and contextuality is to occupy a position that is between the objective pole of scientific theorising and the subjective pole of empathetic understanding'. Nonetheless, on occasions (as in the work of Soja) insightful attempts have been made to combine these different ideas of place as simultaneously physical, mental and social constructions, with a number of writers seeking to develop Henri Lefebvre's notion that places are imagined, perceived and experienced as spaces of representation, representations of space and spaces of social practice at one and the same time (see Lefebvre, 1991; Shields, 1991; Gregory, 1994). Yet Lefebvre's ideas are not straightforward, and revolve around particular understandings of the way the embodied and 'concrete' spaces of everyday life have been commodified and bureaucratized by 'abstract' capitalist processes (see Chapter 2). As a result, collapsing and reinventing the dreaded binary between 'real' and 'imagined' place has proved easier said than done, with the consequence that 'place' remains a term defying easy definition, being used in a variety of disparate ways (Cresswell, 1999). Accordingly, the way human geographers have conceptualized place continues to differ markedly depending on the particular theoretical perspective to which they subscribe.

Here then, we are faced with a bewildering range of possible definitions of 'people' and 'place', each of which is implicated in the perpetuation of a particular perspective on the *relations* between them. For such reasons, Cloke *et al.* (1991) conclude that there is widespread scepticism evident among geographers about the possibility of ever reconciling different approaches to human geography under the banner of one single perspective. Instead, they stress that geographers' acceptance of a diversity of approaches to human geography indicates that they remain mindful of the complexity of the world and are wary about obliterating its variety and heterogeneity by adopting single, 'totalizing' theories which attempt to account for everything. As such, it is perhaps better to refer to the existence of multiple *human geographies* rather than a single human geography, accepting that no one perspective will ever be adequate for making sense of the world. So, even though each perspective we have explored here can be understood partly as emerging from the ones preceding it (i.e. cultural geography imported theories and practices from humanistic geography which imported theories and practices from behavioural geography, etc.), none can be regarded as entirely redundant. On one level, then, we might suggest that each of the perspectives we have considered in this book can illuminate some aspect(s) of the complex relationships between people and place, giving a situated view from particular positions. Nonetheless, Cloke *et al.* (1991) argue that it is important that geographers do not simply wallow in this diversity of approaches, suggesting instead that a healthy and fruitful dialogue should be maintained between proponents of these different perspectives. Inevitably, this dialogue might centre on the acknowledgement that each perspective has particular strengths and weaknesses, being capable of shedding light on some of the relations that exist between people and place (but not others). This type of argument represents a fairly pragmatic way of making sense of the existence of sometimes contradictory ideas about how human beings interact with their environment, stressing that certain perspectives are inevitably better for conducting certain tasks and answering different types of research questions (see also Smith, 1999).

In the rest of this chapter, therefore, we want to think about the choices which geographers make (on a regular basis) about which particular ways of looking at the world are best for answering particular research questions or examining the types of behaviours that occur in particular places. Here, we will suggest that the choice of which perspective they adopt depends on a series of nested questions about the purpose of geographic research; Why are we doing it? Who are we doing it for? How should we communicate our findings? What is useful knowledge? What can we possibly change through our research? These are all complex questions, yet each informs an individual geographer's decision about which perspective is best for them in a particular context. In this chapter we want to consider these briefly under three headings relating to the **philosophy** of human geography, the **morality** of human geography and, finally, the **writing** of human geography.

■ 10.2 Philosophy and human geography

Ultimately, it is important to recognize that questions about which perspective geographers choose to adopt are questions about philosophy. Explicitly or implicitly, all of the perspectives we have reviewed in this book have been informed by (and based upon) particular packages of philosophical thought. These influence the selection of topics for research, the selection of appropriate research methods and the way the research is conducted. In short, the whole endeavour of empirical research – the collection of 'facts' on which geographers base their ideas – is underpinned by philosophical concerns. In simple terms, philosophy (from the Greek for 'love of wisdom') is a discipline which seeks to explore what exists in the world and how we come to know what exists in the world. In Jostein Gaarder's (1996) best-selling novel *Sophie's World*, the (fictitious) philosopher Albert Knox poses two key questions to explain the essence of philosophy to a 14-year-old schoolgirl; 'Who are you?' and 'Where does the world come from?'. Subsequently, he develops a number of analogies, suggesting philosophy is a detective story in which the killer's name can never be revealed (even though there may be many clues) or an attempt to explain the unexplainable:

> A lot of people experience the world with the same incredulity as when a magician suddenly pulls a rabbit out of a hat which has just been shown to them empty. In the case of the rabbit, we know the magician has tricked us. What we would like to know is just how he did it. But when it comes to the world it's somewhat different. We know the world is not all sleight of hand because we are here in it, we are part of it. The only difference between us and the white rabbit is that the rabbit does not realise it is taking part in a magic trick. Unlike us. We feel we are part of something mysterious and we would like to know how it all works . . . . Are you still there Sophie?
>
> (Gaarder, 1996, 13)

Within this academic detective story, different writers and thinkers have evolved specific ideas (or *philosophies*) about the limits of knowledge and existence. These centre on questions of **ontology** (ideas about what actually exists in the world) and **epistemology** (ideas about what we can know and how we come to know it). Consequently, there are many conflicting interpretations of what is knowable and how we might best construct and represent this knowledge.

In sum then, different philosophies seek to define what we can know, what is useful to know and how we can come to know it. Philosophy thus represents abstract ways of thinking which employ forms of logic to organize imaginaries, beliefs and notions of purpose into formal systems of understanding such as positivism, phenomenology, idealism, Marxism and so on (Peet, 1998). While each of these labels in fact covers a wide range of subtly different philosophies, certain geographers have claimed allegiance to one or other of these perspectives over the last 30 years, suggesting that the direction and content of their work has been driven by a particular package of philosophical thought. For example, we have explored in this book how certain philosophical ideas have enjoyed widespread currency among human geographers at particular times, to the extent that it has been common to group geographers together under labels such as behaviouralists, humanists, feminists and so on (see Johnston, 1997). Each group has a particular view on what is actually important in shaping the relationship between people and place and how we might best examine this (e.g. by measuring things that we can observe, by reconstructing the world as others see it or by examining the power relations that underpin society).

To a lesser or greater extent, however, all the aforementioned perspectives can be described as symptomatic of a distinctly **Modernist** conceit that considers it possible to develop an explanatory framework that will lead towards the discovery of the 'truth'. Indeed, academic notions of scientific progress are based on the idea that it is possible to build up knowledge cumulatively into better and better theories (with the ultimate goal of being able to predict and control phenomena). This so-called 'Enlightenment' project is one that dates back at least to the eighteenth century, centring on the idea that it is possible to 'explain' the world and reveal its true order by following certain procedures based on claims of coherence and progress. Like other disciplines, geography has been largely driven by this imperative, introducing new methods of enquiry and philosophies of knowledge in an attempt to produce better, more truthful understandings of the relationship between people and place over time. Yet in more recent years, confidence in this Enlightenment project has wavered for a variety of complex reasons, not least of which has been the impetus of **postmodern** ideas which reject the idea that any one theory will ever be sufficient for explaining all occurrences of a particular phenomena.

Postmodernism is a complex and important idea which eludes straightforward definition. In its widest sense, postmodernism is a term used by cultural commentators, artists and writers to describe any cultural artefact, building or work of art which communicates to different audiences in different ways (rejecting the idea that interpretation is fixed and universal). In a more specific philosophical sense, however, it is a phrase coined by philosophers to indicate dissatisfaction with 'metanarratives' – grand or totalizing theories which attempt to explain away the messiness of the real world. Primarily associated with writers like Jean-Francois Lyotard (1984), postmodernism has come to signify a revolt against the rationality of Modernism and the arrogance implicit in the idea that universal theories are possible (or desirable). Although it has some important connections with post-structuralist philosophies (which question the certainties of established knowledges and languages – see Chapter 9), postmodern philosophy represents a more extreme challenge to the social sciences by dissembling these knowledges in favour of heterogeneity, relativity and (absolute) difference. This challenge is one that poses fundamental epistemological questions about the academic construction of knowledge, rubbishing ideas that accurate description, rational thought or representational accuracy can act as the basis for establishing truth.

Although widely debated in the social sciences throughout the 1980s, geographers were somewhat slow to take the postmodern 'challenge' seriously (Dear, 1988). Primarily, it was through debates on the form of the contemporary city that the ideas and implications of postmodernism first became widely articulated by geographers (see especially Harvey, 1989). Here, postmodernism became widely used as a term which described the type of Western city emerging in the late twentieth century in response to a complex range of economic, social and political reorganizations. Drawing on the work of Fredric Jameson (1984) in particular, Soja (1989) gave a flavour of these changing urban geographies in his book *Postmodern Geographies*, which presented Los Angeles as the quintessential postmodern city. His description of the aesthetic and cultural diversity of downtown Los Angeles began to give some indications as to the new forms of demarcation found in the contemporary city:

> There is a dazzling array of sites in this compartmentalised corona of the inner city: the Vietnamese shops and Hong Kong housing of a redeveloping Chinatown; the Big Tokyo's financed modernisation; the induced pseudo-SoHo of artists' lofts and galleries . . . the strangely anachronistic wholesale markets . . . the capital of urban homelessness in the Skid Row district; the enormous muralled barrio stretching eastwards toward East Los Angeles . . . the intentionally yuppifying South Park redevelopment zone hard by the slightly seedy Convention Center, the revenue-milked towers and fortresses of Bunker Hill.
>
> (Soja, 1989, 239–40)

What Soja appeared to be describing was a *delirious* city permeated by divisions and fractures which could not – in his view – be adequately mapped using the theories and methods utilized by geographers at that time. As a result, while much of his book was couched in the assuredly Modernist vocabularies of Marxist analysis, his interpretation did stress the need for geographers to develop new modes of representation (or 'cognitive maps') to make sense of a city where space–time compression and associated shifts in the mode of production were bequeathing new geographies of inclusion and exclusion.

Equally, questions about the representation and analysis of the countryside began to be posed by rural geographers. For instance, in a review article in the *Journal of Rural Studies* entitled 'Neglected rural geographies: a review', Chris Philo (1992) claimed that rural geographers had tended to explore the countryside through the lens of white, middle-class, masculine and distinctly Modern narratives. His concern was that geographers had primarily attempted to understand and map the countryside by exploring the lifestyles and behaviours of those who constituted the societal mainstream (i.e. white, employed, heterosexual, middle-class men, able of body and of mind). While he conceded that the norm in many rural communities was defined around these co-ordinates, a heightened awareness of 'others' was necessary. As such, he advocated a 'postmodern sensitivity to difference', recognizing that 'social life is . . . fractured along numerous lines of difference constitutive of overlapping and multiple forms of otherness' (Philo, 1992, 200). Following these arguments, the 1990s have witnessed a series of studies of rural 'others' such as women, children, teenagers, the elderly, gays, lesbians, blacks and ethnic minorities, the unemployed and poor, gypsies, 'New Age travellers' and people living in various alternative lifestyle communities (see particularly Cloke and Little, 1997, and Milbourne, 1997). More importantly, perhaps, this postmodern turn has involved a deconstruction of existing academic notions of the 'rural' in favour of a sustained exploration of how notions of rurality are invoked and revoked by different groups (Phillips, 1998). Accordingly, the distinction between the representation and reality of the countryside has appeared difficult to sustain as the rural takes on meanings which cannot be rooted in any one person's experience (Thrift, 1994).

Here, then, both urban and rural geographers are engaging with complex postmodern debates about how academic knowledge claims to represent the 'real' in some way. For instance, Jean Baudrillard's (1983) insistence that we cannot ever hope to capture and communicate 'reality' through signs and symbols has been explored by some geographers. His notion that signs and concepts can exist free from any referent (e.g. that words can have meanings in themselves, independent of any 'reality' outside of themselves) suggests that we exist in a 'general economy of signs' where no one text can claim to be more truthful or authentic than any other. Some practitioners have therefore stressed the need for geographers to abandon any hope of reflecting 'reality' in an organized, premeditated or rational manner, instead proposing that they 'make space' for difference and otherness by refraining from categorizing, explaining or interpreting things that happen in the world. In the words of Marcus Doel (1999, 198), for example, it is 'ill-mannered' to impose consistency on specific events, but ethical to let 'space take place', respecting their (pure) difference and holding off from obliterating their specificity through attempting to generalize (see also Doel and Clarke, 1994). Here, Modern ideas and theories are cast into the maelstrom of uncertainty to be replaced by disjuncture, deformation and deconstruction. Such a view is (more or less) consistent with Lyotard's argument that meta-theory converts difference into sameness, in effect 'flattening' the complexity of social life by suggesting there are general 'scripts' by which life is written. Perhaps you can start thinking here about how the academic perspectives we have looked at in this book ignore some of the things that might matter to certain people in their everyday lives. For example, superstition, astrology, premonition, fate, reincarnation and religion have rarely been considered by geographers as shaping human actions (being deemed totally irrational and hence unexplainable), yet it is entirely possible that people believe in some or all of these things in a manner that cannot be captured in any generalized account of human spatial behaviour.

Whether one prefers to subscribe to a postmodern world of variation or a Modern one of predictability depends on one's reading of philosophical debates. Although many geographers do not explicitly refer to these debates, we cannot simply dismiss them as irrelevant to the practices of 'doing' human geography, as different views about what exists in the world have influenced the way that geographers have chosen their objects and methods of study in a profound manner. We do not intend to dwell further on these issues here, but hopefully it has become obvious that all human geography is influenced by ideas derived from philosophy. As geographers, therefore, we are all personally obliged to think through these questions of ontology and epistemology and decide what particular package(s) of philosophical thought we subscribe to. Ultimately, this decision is entwined with other key questions about the purpose and goals of academic research, forcing us to think about our senses of right and wrong, worth and worthlessness. It is to these questions of morality that we turn to next.

Reading

There are many books which purport to explain how different philosophical ideas have been interpreted and implemented by human geographers. Ron Johnston's (1997) *Geography and Geographers* is organized chronologically using the concept of paradigms – particular frameworks of explanation which have enjoyed widespread currency among Anglo-American geographers at particular times. Now in its fifth edition, it presents a seemingly definitive story of human geography's intellectual development – though you might want to contrast it with David Livingstone's (1992) *The Geographical Tradition*. Although this scarcely mentions developments over the last 30 years, it does highlight the problems inherent in organizing the 'situated messiness' of human geography using periodizing concepts like paradigms. Other key introductions to philosophy and human geography include Cloke *et al.* (1991), Unwin (1992) and Peet (1998); it might be a good idea to re-read the various chapters in this book in parallel with any one of these accounts of the development of geographic thought. For a good, general introduction to postmodernism, see Smart (1993).

10.3 Moral geographies, immoral geographers?

In Chapter 8 we began to describe one aspect of geography's engagement with questions of morality. Here, 'moral geography' was defined as being concerned with interrogating how different senses of right and wrong are played out in different places and how this bequeaths places and spaces where behaviours routinely acceptable elsewhere are not tolerated (Smith, 1997a). Yet geography's interest in questions of morality is somewhat wider than this, simultaneously exploring the morality of the discipline itself and questioning the values of its practitioners. For example, given that academic research is in part the product of personal experience, morality and politics (Rose, 1993), does academic positionality and privilege preclude geographers from making meaningful contributions to debates about poverty and exclusion? Even with the adoption of seemingly sensitive forms of enquiry, designed to promote empathetic understanding of different social experiences, is geographical research always in danger of exploiting the people and places it studies? And is it possible, or even desirable, to try to change the lives of the marginalized and oppressed groups through our research? In short, can geographical research be made a force for good?

These are not new concerns, being raised periodically by geographers concerned with the trajectory of the discipline and its ability to ameliorate social, economic and environmental problems. For instance, reflecting on the contribution of quantification to geographical scholarship, a number of writers in the early 1970s argued vociferously for a more 'relevant' geography (e.g. Berry, 1972; Harvey, 1973; Smith,

1975). For these writers, geography at the time appeared to be populated by practitioners who were constructing models and theories in splendid ignorance of the problems of those living in the world beyond the 'ivory towers' of academia. In the frequently cited words of David Harvey:

> There is a clear disparity between the sophisticated theoretical and methodological frameworks which we have developed and our ability to say anything really meaningful about events as they unfold around us . . . . There is an ecological problem, an urban problem, a debt problem, yet we seem incapable of saying anything in depth or profundity about any of them.
>
> (Harvey, 1973, 129)

While Harvey was dedicated to a 'radical' Marxist critique of the inequalities between different social classes, others had differing opinions as to how geography could best contribute to the solution of social problems. Some, for example, felt that geographers needed to be working more closely with policy-makers, planners and governments, involved in helping to decide how resources could be more equitably distributed between different people and places (Berry, 1972). Against this, some articulated a concern that geographers should aim not to become merely part of the decision-making apparatus but should find other ways of creating a more just world via teaching and research (Chisholm, 1971). Yet others remained highly sceptical of geographers' ability to change anything, and continued to argue for the production of knowledge for its own sake.

Yet in the 20 years or so following this debate, the idea that human geography can in some way act as a positive force for social change, resulting in a more just and equal world, remained an important motivation for many geographers. Most recently, for example, many English-language geographers (and others beyond the Anglo-American diaspora) have expressed a commitment to the pursuit of '**critical geography**', organizing Internet discussion groups, conferences and publications dedicated to critical research and teaching (see Kitchin and Hubbard, 1999). While such a trajectory has clearly been subject to contestation, the idea that geography should be 'critical' thus represents an important rallying point in contemporary geographical debate. However, while this commitment to critical geographic teaching and research currently appears a widely shared aspiration, geographers attribute many different meanings to the term 'critical'. It is impossible to define what critical geography might stand for with any precision, although Chouinard and Grant (1996) have argued that the central characteristic of critical geography is its emphasis on studying how multiple processes of oppression and exclusion (sexism, racism, disabilism, etc.) intertwine to construct the unequal social, economic and political relationships which exist in different places. In so doing, they draw attention to the continuities existing between contemporary critical geography and the 'radical' geographies of the 1970s, albeit recognizing that the scope of current research has expanded beyond the prioritization of class relations. For them what is most distinctive about contemporary critical geography is that it *actively* seeks to destabilize and undermine processes of oppression both within and without the academy (see also Short and Lowe, 1990; Castree, 1999).

The espousal of a critical stance in human geography thus raises intriguing possibilities for research, suggesting that those involved in geographical scholarship have not only a right, but a duty, to be involved in social, economic and political change. Yet with the post-structural agenda problematizing the relationship between researcher and subject of research (see Chapter 9), such ambitions have simultaneously raised important epistemological questions concerning the methods and politics of research. Procter (1998) suggests that ethical dilemmas therefore emerge at every stage of the research process, and that the way that we employ particular research techniques, construct knowledge and disseminate this knowledge involves consideration of a range of moral issues. This can perhaps be illustrated by way of an example. Noting the deficit of research into the geographies of homosexuality (though see Chapter 9), Kim England (1994) set out to explore lesbians' use of space in a Canadian city. Immediately, she faced a number of problems, in that the lesbian community in this particular city constituted a largely 'hidden' population. This meant that she had to think about how she would access this group, and whether she would adopt an overt or covert role (i.e. whether she would explain that she was involved in doing a research project or not). She then had to decide how she would collect 'data' from these women while respecting their own ability to place themselves in theoretical contexts. Finally, she had to think about how she could write her results up in a way that protected the confidentiality and anonymity of her respondents. Here, she was aware that the lesbian community was one that chose to be 'invisible' to avoid prejudice and hostility. Mindful of Cindi Katz's (1994, 71) suggestion that 'there is a danger in making the practices of the oppressed visible to those who dominate', her ultimate decision was to abandon her research altogether.

While few geographers would perhaps go to this extreme to avoid doing harm to their research respondents (e.g. Valentine's 1993 study of lesbian time–space strategies claims to be based on interviews conducted in the fictitious town of 'Melchester'), the idea that geographers can inflict emotional (and even physical) violence on the people that they study is one that is taken seriously. As Sidaway (1992, 407) contends, research of 'other' people and places can counter the narrow-mindedness that can sap the vitality of geographical enquiry, but can also be patronizing, exploitative or expose marginal populations to greater risk of harm. Accordingly, **reflexivity** (reflecting on ourselves in relation to other people and other things) has certainly become a central concern for human geographers as they think about the result their research will have on themselves and others (Duncan, 1993). On one level, this involves a careful consideration as to whether our research causes physical or mental stress for participants (e.g. by invading their privacy), while on another it involves consideration of whether we are prepared to work for particular organizations, publish in particular journals or make our research findings available to certain groups. However, in an era where many colleges and universities are preoccupied with quantity of research output (a further symptom of the

McDonaldization described in Chapter 2, perhaps), it is unsurprising that many academics feel pressurized to seek research funding from organizations that they do not feel entirely happy working with (Sidaway, 1997).

Again, these moral questions entwine with philosophical issues of epistemology and ontology, particularly those centring on postmodernism. For example, in an unreservedly Modernist reading of geographical ideas, Peet (1998) condemns the postmodern excesses of some writers for perpetuating an 'amoral geography'. For him, the 'scrumpled' geography of (in)difference and disjuncture proposed by Marcus Doel (1999) is symptomatic of the way geographers can do damage to oppressed people by neglecting to compare their situation with those of other (more affluent) people. Peet (1998, 242) claims that Doel's geography is 'an invitation to personal indulgence reminiscent of those days when academics had triple-barrelled names and were upper-class dilettantes researching quaint topics of pure esoterica' (though against this, Doel might claim that his is a distinctly moral geography that respects the distinctiveness of particular events, with morality and rights being specific to the event). Elsewhere, Peet berates the tendency for other geographers to 'read' space through other people's writing, suggesting this creates a 'dense intertextuality which precludes contact with the world outside the fetishistic dream' (Peet, 1998, 226). Equally, Neil Smith's (1996a) parody 'Rethinking sleep' (which appeared in a journal which has acted as something of a showcase for postmodern geographical writing) attempts to show up the work of some geographers as self-indulgent and nihilistic by ironically proclaiming the 'resistive potential' of doing nothing. In a similar vein, Pacione holds on to a relatively straightforward notion of what is 'useful' knowledge:

> It is clearly a matter of individual conscience whether geographers study such topics as the iconography of landscape or the optimum location for health centres, but the principle underlying the kind of useful geography espoused by most applied geographers is a commitment to improving social, economic and environmental conditions. There can be no compromise, no academic fudge; some academic research is more useful than other work.
>
> (Pacione, 1999, 4)

Whether such arguments hold in the light of postmodern critiques of universal truth is a moot point – after all, it is difficult to imagine that a change in social, economic or environmental conditions can benefit all groups equally, or that all groups will ever agree on what a correct distribution of goods across society should be.

These debates rage on as geographers seek to mark out the contours of a moral and ethical geography. At the current time, however, there are a number of geographers who are trying to resolve the tension between Modern and postmodern senses of morality by creating a geography that reaches out to marginalized people (and places) without heroically assuming the duty of assessing their lives against transcendental yardsticks of 'right/wrong' and 'good/bad' that may have little relevance for them. For instance, when Derek Gregory (1994, 416) writes of a critical human geography that reaches out from one body to another, 'not in a mood of arrogance, aggression and conquest, but in a spirit of humility, understanding and care', he perhaps captures a mood within contemporary human geography that it is necessary to acknowledge certain moral absolutes based on a recognition of shared human traits (e.g. biological needs of shelter, food and water; freedom of speech; avoidance of harm; etc.). While geography is never likely to copy a discipline like psychology (where researchers are obliged to observe strict ethical codes) we can begin to see the development of some moral 'guiding principles' here which might usefully serve to inform geographical research of people and place (see also Hay, 1995; Smith, 1997b). These guiding principles would concern not just the way research is conducted, but also the way it is 'written' and disseminated. It is these concerns with which we conclude this chapter.

■ Reading

David M. Smith's work over the last 30 years represents a (more or less) coherent attempt to encourage geographers to engage with questions of human welfare, equality and need. His *Geography and Social Justice* (1995) represents perhaps the most sustained engagement by a geographer into questions of social justice. Chapters 2 and 3 are particularly recommended as an introduction to key positions in moral philosophy such as egalitarianism, liberalism and social contractualism, while his later chapters show how these concepts might apply to 'problems' of inequality. Smith's *Moral Geographies* (2000) is an accessible and interesting overview of the subject area. David Harvey's work over a similar time-span has perhaps taken more convoluted twists and turns – taking in geographic readings of Marx *(The Limits to Capital)* as well as disquisitions on the nature of the urban experience *(Consciousness and the Urban Experience)* – but questions of social justice have never been far beneath the surface. His essay in *Antipode* (a journal of 'radical' geography) entitled 'Postmodern morality plays' (1992) is a useful reflection on many of the debates touched on in this section.

■ 10.4 Doing geography: telling stories

While philosophical introspection and debate characterize contemporary geography, few would disagree that the ultimate task of academic geography has changed little over the years – in short, that it is concerned with the production, discussion and dissemination of knowledge. Here we should take time to reflect on the actual techniques used to create that knowledge, remembering that research is never simply an abstract process of data collection. By this, we mean that research is inherently political, in that the act of gathering data involves the researcher negotiating relationships with those that they are seeking to gather data from (or about).

As with any other social relationship, the type of dynamic played out between researcher and researched involves a complex power relation, with the researcher often in a situation where there is potential for them to 'lead' the respondent towards particular types of responses. Many of the debates about the merits of quantitative research techniques (e.g. questionnaires, experiments, surveys) versus qualitative research techniques (e.g. observation, interviews, discourse analysis) have accordingly focused on questions of **bias**. Assertions that either quantitative or qualitative techniques are more objective (i.e. less prone to bias) are ultimately difficult to sustain, although in some situations geographers have preferred to use more in-depth, intensive research methods in an attempt to develop a more balanced relationship between researcher and researched. As we saw in Chapter 9, some have argued that these alternatives to 'traditional' quantitative techniques are most appropriate when geographers are seeking to empathize with some of those groups whose voices have often been marginalized (such as gay groups, people with disabilities, women, people of colour, the young and the elderly, the homeless, etc.). However, this does not imply that quantitative techniques are never useful for studying the geographies of these groups. Instead, drawing on the debates about ethics we considered above, most geographers would concede that choice of research technique relies on a careful consideration of **positionality** (i.e. an awareness of how our personal characteristics influence the way we 'ask' questions and 'listen' to responses).

Consequently, the methods and techniques that might be appropriate for one geographer researching a given situation might be manifestly inappropriate for another. Returning to the geographies of gender, for example, it is widely accepted that female geographers might be able to ask different questions of women than a male geographer can (Sparke, 1996). Equally, the type of answers they receive might be very different than those which a male geographer might receive. While this does not mean that only women can study other women (or men study other men), it does suggest that geographers need to remain wary of their status as insiders/outsiders, changing their preferred mode of research to fit the circumstances of a specific research situation. Such concerns about positionality of course extend into other stages of the research process, including analysis, interpretation and, ultimately, publication. The last is of major importance, as it is by publishing research findings in books, conference papers, journal articles, posters, maps and diagrams that geographers hope to disseminate their ideas to those within the academy (other researchers, teachers and students) as well as outside it (policy-makers, user groups, the 'public' and so on). As we explored in Chapter 7, geographers are always engaged in the creation of place through the process of representation, in the sense that places do not come ready-defined and labelled (pre-packed) – it is geographers who decide how to define and represent places, not the places themselves. Because of this, and in the light of recent post-structural and postmodern debates about the construction of knowledge, many geographers have argued for a greater awareness of the importance of geographical representation and 'writing' (Barnes and Duncan, 1992; Gregory, 1994; McNeill, 1998). In this context, writing is clearly about more than just engaging the reader through carefully constructed, lucid and thoroughly referenced prose; it is about the *construction of meaning* and the *production of place*:

> Although we usually think about writing as a mode of 'telling' about the social world, writing is not just a mopping-up activity at the end of a research project. Writing is also a way of 'knowing' – a method of discovery and analysis. By writing in different ways, we discover new aspects of our topic and our relationship to it. Form and content are inseparable.
>
> (Richardson, 1994, 516)

The implication here is that certain ways of writing encompass certain ways of thinking about the world, prioritizing some things, leaving others out. The choice of a way of writing is therefore an inherently political (and *moral*) choice.

To explain this, it is perhaps useful to turn to the **hermeneutic** ideas developed by Duncan and Ley (1993) amongst others. This idea refers to the way that the meaning of texts can only be understood in relation to other texts. In Figure 10.1, any 'text' produced by a geographer is taken to be a representation of the **extratextual** geographical world, and as such, it can only ever be a partial representation (a partial or situated truth) of what is considered to exist in the world. The question of *how* it distorts the extratextual world depends on how the text relates to the **intratextual** world – i.e. the other books, theories, ideas and authors which inform the construction of the text. For instance, consider writing about the geography of a university or college. Clearly, this is a complex physical and social setting, and there is no way that it is possible to represent it in totality. (The only way this would be possible would be to create a full-size replica! And even then, how would the complexity of social relations and cultural forms be replicated?) As such, it is necessary to decide which particular 'story' (or geography) of the university you want to tell. This choice will be based not only on certain moral and philosophical principles (as already discussed in this chapter) but also by the type of ideas and theories which appear to be relevant or interesting. It may be that one has been reading about the effect of classroom design on student attainment, in which case it might be written up as a 'scientific' study of environmental psychology (see Chapter 3); alternatively, it may be that the study is influenced by Foucault's notion of disciplinary space, in which case it might be written as an examination of knowledge–power relations in the lecture theatre (see Chapter 8).

Figure 10.1 Textual relations

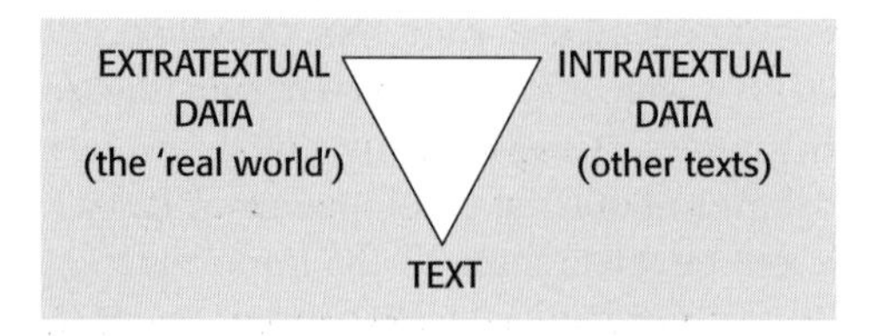

Source: Based on Duncan and Ley (1993)

Rejecting the notion of mimesis (i.e. that it is possible to reflect the reality of the world; see Chapter 7), we are therefore faced with a situation where meaning is produced from text to text, with geographers creating 'new worlds out of old texts' (Barnes and Duncan, 1992, 3). This means that geographers are now rarely innocent about the way they use particular writers, ideas and perspectives to create representations, being aware that their writings are exercises of discursive definition and power which serve to define place itself. Traditionally, however, most geographers have unquestioningly adopted a positivist scientific way of writing which presupposes some notion of distance, being framed in a disembodied, male 'academic' voice (writing in the third person) which emphasizes predictability and quantification. This voice appears to promise that any researcher would have produced the same account if he/she had followed the same 'objective' procedures. As we saw in Chapter 1, much geographic writing is in this genre, depicting human beings using the language of spatial science – as predictable actants moving on a mappable and knowable surface. In subsequent years, geographers have come to realize that they cannot represent the 'truth' of the relations between people and place, but merely tell a story of this relationship written from their own particular perspective. This has encouraged them to explore the potential of new and often innovative ways of writing that acknowledge the positionality of the author.

So although many geographers still write using a scientific **trope** (or style of writing), it is possible to see geographers adopting tropes such as travel writing, journalism and autobiography. Here, the writings of authors as diverse as Walter Benjamin, Helen Cixous, Patrick Wright, Beatrix Campbell, Iain Sinclair, Guy Debord and Mike Davis have all been cited as blurring the boundaries between practice and discourse (or life and imagination) in innovative ways. For instance, Mike Davis's *City of Quartz* (1991) has been proclaimed as a 'plucky, streetwise' and eminently readable text that conveys how Los Angeles is being torn apart by violence and disorder (Soja, 1996, 27). The fact that Davis admits to exaggerating for effect, making spurious reference to statistical sources and changing the details of particular encounters he has in the city seems to matter not – for Derek Gregory (1994, 305) he succeeds in 'making the city come alive'. He seems to do this by engaging with the street-level activity of the city, favouring a view from the 'street' rather than an abstract view from afar. As such, rather than claiming some sort of separate Archimedean point from which the world can be critiqued at a distance (as was the goal of spatial science), Davis acknowledges his story is the product of a particular set of highly individual encounters with the city and its people. Telling stories of the street has a long track record in both anthropology and urban sociology – certainly back to Walter Benjamin's *One Way Street* (1925–26), which used his wanderings on the streets as a device on which to hang a series of reflections on everyday life – and this seems to offer one way in which geographers can begin to write themselves into their texts.

Here there are important connections with the practices of **ethnography** (literally, writing about a way of life). Influential on humanistic geographers (see Chapter 4), the ethnographic technique is an empathetic and naturalistic approach where the geographer seeks to engage with people in their everyday life-world rather than creating an artificial research situation (an experiment, interview or questionnaire). This can involve both observational techniques and informal conversations, often focused on a very small area or limited number of people over a long period of time (creating a 'thick description' of the interactions that occur between people and place). But, taking into account the hermeneutic that exists between the text and the outside world, it is evident that those who seek to describe a place using ethnography can never do so in a mimetic way – revealing its 'truth'. As the feminist theorist Donna Haraway contends:

> There is no unmediated photograph or passive camera obscura in scientific accounts . . . there are only highly specific visual possibilities, each with a wonderfully detailed, active, partial way of organising worlds. All these pictures of the world should not be allegories of infinite mobility and interchangeability but of elaborate specificity and difference and the loving care people might take to learn how to see faithfully from another's point of view.
>
> (Haraway, 1991, 190)

As a result, some geographers support the idea that texts should be open to multiple voices (e.g. Women and Geography Study Group, 1997). Here, the idea is that the people we study should be able to represent *themselves* using their *own* words (or photographs, pictures and poems), and this offers an interesting way of breaking down the power relations that exist between researcher, research community and the readers of a text. To a certain extent, this demands that the researcher abandons her/his authority over the text, but might provide an opportunity for marginalized and oppressed people to tell their own stories.

Finally then, we must again recognize there are no rights and wrongs here – and certainly no definitive guidebooks that can tell us how to research, interpret and write human geographies. As this final chapter has hinted, geographers seldom agree on the right way of doing things. While some no doubt find this frustrating, for others it is to geography's credit that its practitioners spend time interrogating what they do (and how they do it). You will no doubt have your own view on this. Perhaps, like us, you find these debates liberating and exciting and want to spend some time thinking through the possible implications of doing human geography in particular ways. On the other hand, perhaps you find all this naval-gazing somewhat tiresome, and think that geographers need to stop worrying so much about what they do, and get on and do it! Whatever your opinion, hopefully you can see that the ideas about people and place that we have explored throughout the book are not just of 'academic' interest, but have repercussions for society in a variety of complex ways. As we have shown, ideas developed by geographers in one way or another have influenced the way that everyday settings like houses, restaurants, libraries, universities, parks, gyms, sports centres, nightclubs, supermarkets, streets, airports, gardens and stadia are designed and used. Equally, we have suggested that taken-for-granted distinctions between near and far, core and periphery, town and country, north and south, or local and foreign are inevitably informed by geographical concepts and representations. As such, geographic ideas may simply be stories or representations of the relationships that exist between people and place, but on the basis of these stories, people have had their lives changed – sometimes for the better, sometimes for the worse. While we can never hope to predict how the knowledge we construct will be used (and by whom), thinking through some of the questions in this chapter might help you produce stories which are respectful of the delicate and complex relationships existing between people and places.

■ Reading

Questions of how geographers decide to adopt particular ways of writing geography are broached in Barnes and Duncan (1992) and Duncan and Ley (1993), while the ideas about writing the streets presented here are discussed in McNeill (1998). A more wide-ranging (and ambitious) text on the connections between geographic representation, knowledge and practice is Derek Gregory's (1994) *Geographical Imaginations* – a difficult but ultimately rewarding read. Finally, we might suggest that the best way of thinking through the practical, ethical and philosophical issues which face human geographers is by 'doing' geography. Good, recent texts on research methods and techniques include Flowerdew and Martin (1997) and Kitchin and Tate (2000), while Chapter 14 of Shurmer-Smith and Hannam (1994) outlines some of the different ways it is possible to 'do' cultural geography. The use of ethnographic techniques in geography is explored in Cook and Crang (1995) and Hughes *et al.* (2000). We hope they might provide some ideas as to how you could explore some of the theories and ideas contained in this book in the places that you encounter on an everyday basis.

bibliography

Adler S. and Brenner J. (1992) 'Gender and space: lesbians and gay men in the city' *International Journal of Urban and Regional Research* 16: 24–31

Agnew J., Livingstone D. and Rogers A. (eds) (1996) *Human Geography: An Essential Anthology* Oxford: Basil Blackwell

Aitken S. (1992a) 'Person–environment theories in contemporary perceptual and behavioural geography I: personality, attitudinal and spatial choice theories' *Progress in Human Geography* 15: 179–93

Aitken S. (1992b) 'Person–environment theories in contemporary perceptual and behavioural geography II: the influence of ecological, environmental learning, society/structural, transactional and transformational theories' *Progress in Human Geography* 16: 553–62

Aitken S. (1998) *Family Fantasies and Community Space* New Brunswick: Rutgers University Press

Alderman D. H. (1997) 'TV news and the representation of place: observations on the O. J. Simpson case' *Geografiska Annaler* 79B: 83–95

Allen J. (1995) 'Crossing boundaries: footloose multinationals?' in Allen J. and Hamnett C. (eds) *A Shrinking World?* Oxford: Open University

Allen J. (1999) 'Worlds within cities' in Massey D., Allen J. and Pile S. (eds) *City Worlds* London: Routledge

Allen J. and Hamnett C. (1995) *A Shrinking World?* Oxford: Open University

Anderson K. (1995) 'Culture and nature at the Adelaide Zoo: at the frontiers in "human" geography' *Transactions, Institute of British Geographers* 20: 275–94

Anderson K. (1998) 'Animals, science and spectacle in the city' in Wolch J. and Emel J. (eds) *Animal Geographies: Place, Politics and Identity in the Nature–Culture Borderlands* London: Verso

Appadurai A. (1990) 'Disjuncture and difference in the global cultural economy' *Theory, Culture, Society* 7: 295–310

Atkinson D. (1998) 'Totalitarianism and the street in fascist Rome' in Fyfe N. (ed) *Images of the Street* London: Routledge

Augé M. (1995) *Non-places: Introduction to an Anthropology of Supermodernity* London: Verso

Austen J. (1946) [1813] *Pride and Prejudice* London: Zodiac Press

Bachelard M. (1958) *The Poetics of Space* Boston, MA: Beacon Press

Baldry C. (1999) 'Space: the final frontier' *Sociology* 33: 535–53

Baldwin E., Longhurst B., McCracken S., Ogborn M. and Smith G. (1999) *Introducing Cultural Studies* London: Prentice Hall

Barker R. (1968) *Ecological Psychology: Concepts and Methods for Studying the Environment of Human Behaviour* Stanford, CA: Stanford University Press

Barnes T. and Duncan J. (1992) *Writing Worlds: Discourse, Text and Metaphor in the Representation of Landscape* London: Routledge

Barnes T. and Gregory D. (1997) 'Place and landscape' in Barnes T. and Gregory D. (eds) *Reading Human Geography: The Poetics and Politics of Inquiry* London: Edward Arnold

Barrell J. (1980) *The Dark Side of the Landscape* Cambridge: Cambridge University Press

Bassett C. (1997) 'Virtually gendered' in Gelder K. and Thornton S. (eds) *The Subcultures Reader* London: Routledge

Baudrillard J. (1983) *Simulations* New York: Semiotext(e)

Bechtel R. B. (1997) *Environment and Behavior: An Introduction* London: Sage Publications

Bell D. (1992) 'Insignificant others: lesbian and gay geographies' *Area* 23: 323–9

Bell D. (1995) 'Pleasure and danger: the paradoxical spaces of sexual citizenship' *Political Geography* 14: 139–53

Bell D. (1997) 'The anti-idyll: rural horror' in Cloke P. and Little J. (eds) *Contested Countryside Cultures* London: Routledge

Bell D. and Valentine G. (1995a) 'Queer country: rural lesbian and gay lives' *Journal of Rural Studies* 11: 113–22

Bell D. and Valentine G. (eds) (1995b) *Mapping Desire: Geographies of Sexualities* London: Routledge

Bell D. and Valentine G. (1997) *Consuming Geographies: You Are Where You Eat* London: Routledge

Benko G. (1997) 'Introduction: modernity, postmodernity and the social sciences' in Benko G. and Strohmayer U. (eds) *Space and Social Theory: Interpreting Modernity and Postmodernity* Oxford: Blackwell

Bermingham A. (1986) *Landscape and Ideology: The English Rustic Tradition 1740–1860* London: Thames & Hudson

Berry B. (1972) 'More on relevance and policy analysis' *Area* 4: 36–40

Bettelheim B. (1976) *The Uses of Enchantment: the Meaning and Importance of Fairy Tales* London: Thames and Hudson

Benjamin W. (1979) [1925/1926] *One Way Street* London: New Left Books

Bey H. (1991) *Territorial Autonomous Zones* New York: Autonomedia

Binnie J. and Valentine G. (1999) 'Geographies of sexuality – a review of progress' *Progress in Human Geography* 23: 175–87

Bonnes M. and Secchiaroli G. (1995) *Environmental Psychology: A Psycho-social Introduction* London: Sage

Borden I. (1996) 'Beneath the pavement, the beach: skateboarding, architecture and the urban realm' in Borden I., Kerr J., Pivario A. and Rendell J. (eds) *Strangely Familiar: Narratives of Architecture in the City* London: Routledge

Bordo S. (1993) *Unbearable Weight: Feminism, Western Culture and the Body* Berkeley, CA: University of California Press

Bourdieu P. (1988) *Homo Academicus* Cambridge: Polity Press

Boyle M. and Hughes L. (1995) 'The politics of urban entrepreneurialism in Glasgow' *Geoforum* 15: 1459–69

Brantingham P. J. and Brantingham P. L. (1992) 'Nodes, paths and edges: considerations of the complexity of crime and the physical environment' *Journal of Environmental Psychology* 13: 3–29

Bunce M. (1994) *The Countryside Ideal* Routledge: London

Burgess J. (1985) 'News from nowhere: the press, riots and the myth of the inner city' in Burgess J. and Gold J. R. (eds) *Geography, the Media and Popular Culture* London: Croom Helm

Burton R. (1964) [1893] *Personal Narrative of a Pilgrimage to Al-Madinah and Meccah*, mem. ed, New York: Dover

Butler J. (1990) *Gender Trouble: Feminism and the Subversion of Identity* New York: Routledge

Butler R. (1999) 'The body' in Cloke P., Crang, P. and Goodwin M. (eds) *Introducing Human Geographies* London: Edward Arnold

Butler R. and Bowlby S. (1997) 'Bodies and spaces: an exploration of disabled people's experiences of public space' *Environment and Planning D: Society and Space* 15: 411–33

Buttimer A. (1976) 'Grasping the dynamism of the lifeworld' *Annals, Association of American Geographers* 66: 277–92

Canter D. V. (1977) *The Psychology of Place* London: Architectural Press

Canter D. V. (1991) 'Understanding, assessing and acting in places: is an integrative framework possible?' in Garling T. and Evans G. W. (eds) *Environment, Cognition and Action: An Integrated Approach* Oxford: Oxford University Press

Cassidy T. (1997) *Environmental Psychology: Behaviour and Experience in Context* London: Psychology Press

Castree N. (1999) 'Out there? In here? Domesticating critical geography' *Area* 31: 81–6

Chaney D. (1994) *The Cultural Turn: Scene-setting Essays in Contemporary Social History* London: Routledge

Chapman T. (1999) 'Spoiled home identities: the experience of burglary' in Chapman T. and Hockey J. (eds) *Rethinking the Domestic* London: Routledge

Chisholm M. (1971) 'Geography and the question of relevance' *Area* 3: 28–32

Chouinard V. and Grant A. (1996) 'On being not even anywhere near "The Project": ways of putting ourselves into the picture' in Duncan N. (ed) *Bodyspace: Destabilising Geographies of Gender and Sexuality* London: Routledge

Clark D. (1997) *Urban World/Global City* London: Routledge

Cloke P. (1999) 'Self-other' in Cloke P., Crang, P. and Goodwin M. (eds) *Introducing Human Geographies* London: Arnold

Cloke P. and Little J. (eds) (1997) *Contested Countryside Cultures* London: Routledge

Cloke P. and Perkins H. (1998) '"Cracking the Canyon with the awesome foursome": representations of adventure tourism in New Zealand' *Environment and Planning D: Society and Space* 16: 185–218

Cloke P., Philo C. and Sadler D. (1991) *Approaching Human Geography: An Introduction to Current Theoretical Debates* London: PCP Press

Cohen S. (1995) 'Sounding out the city: music and the sensuous reproduction of place' *Transactions, Institute of British Geographers* 20: 434–46

Cook I. and Crang M. (1995) *Doing Ethnographies* Norwich: Geobooks

Coppock J. and Duffield B. (1975) *Recreation in the Countryside: A Spatial Analysis* London: Macmillan

Corbin A. (1994) *The Foul and the Fragrant* Cambridge, MA: Harvard University Press

Cosgrove D. (1985) 'Prospect, perspective and the evolution of the landscape idea' *Transactions, Institute of British Geographers* 10: 45–62

Cosgrove D. (1989) 'Geography is everywhere: culture and symbolism in human landscapes' in Gregory D. and Walford R. (eds) *Horizons in Human Geography* London: Macmillan

Cosgrove D. and Daniels S. (1988) *The Iconography of Landscape* Cambridge: Cambridge University Press

Cox G. (1993) '"Shooting a line": field sports and access struggles in Britain' *Journal of Rural Studies* 9: 267–76

Cox G. and Winter M. (1997) 'The beleaguered "other": hunt followers in the countryside' in Milbourne P. (ed) *Revealing Rural Others: Representation, Power and Identity in the British Countryside* London: Pinter

Cox K. R. (1981) 'Bourgeois thought and the behavioural geography debate' in Cox K. and Golledge R. G. (eds) *Behavioural Problems in Geography Revisited* London: Methuen

Crang M. (1998) *Cultural Geography* London: Routledge

Crang P. (1999) 'Local–global' in Cloke P., Crang P. and Goodwin M. (eds) *Introducing Human Geographies* London: Edward Arnold

Cresswell T. (1992) 'The crucial where of graffiti: a geographical analysis of reaction to graffiti in New York' *Environment and Planning D: Society and Space* 10: 329–44

Cresswell T. (1996) *In Place/Out of Place: Geography, Ideology and Transgression* Minneapolis: University of Minnesota Press

Cresswell T. (1997) 'Weeds, plagues and bodily secretions: a geographic interpretation of metaphors of displacement' *Annals of the Association of American Geographers* 87: 330–45

Cresswell T. (1999) 'Place' in Cloke P., Crang, P. and Goodwin M. (eds) *Introducing Human Geographies* London: Edward Arnold

Crouch D. (1998) 'The street in the making of popular geographic knowledge' in Fyfe N. (ed) *Images of the Street: Planning, Identity and Control in Public Space* London: Routledge

Crouch D. and Matless D. (1996) 'Refiguring geography: parish maps of common ground' *Transactions, Institute of British Geographers* 21: 236–55

Csikszentmihalyi M. and Rochberg-Halton E. (1981) *The Meaning of Things: Symbols and the Self* Cambridge: Cambridge University Press

Cunningham J. (1997) 'The trouble with Boothby Graffoe' *Guardian* 12 August: 7

Daly G. (1998) 'Homelessness and the street: observations from Britain, Canada and the United States' in Fyfe N. (ed) *Images of the Street: Planning, Identity and Control in Public Space* London: Routledge

Dandeker C. (1990) *Surveillance, Power and Modernity: Bureaucracy and Discipline from 1700 to Present* Cambridge: Polity Press

Daniels S (1985) 'Arguments for a humanistic geography' in Johnston R. J. (ed) *The Future of Geography* London: Methuen

Daniels S. (1993) *Fields of Vision: Landscape Imagery and National Identity in England and the United States* Cambridge: Polity Press

Davis M. (1991) *City of Quartz: Excavating the City of the Future in Los Angeles* London: Verso

Davis M. (1998) *The Ecology of Fear: Los Angeles and the Imagining of Disaster* New York: Metropolitan

de Certeau M. (1984) *The Practice of Everyday Life* Davis, CA: University of California Press

Dear M. (1988) 'The post-modern challenge: reconstructing human geography' *Transactions, Institute of British Geographers* 13: 262–74

Dear M. and Wolch J. (1989) 'How territory shapes social life' in Wolch J. and Dear M. (eds) *The Power of Geography* Winchester, MA: Unwin Hyman

Debord G. (1967) *The Society of the Spectacle* London: Rebel Press

Deleuze G. and Guattari F. (1988) *A Thousand Plateaus* London: Athlone Press

Doel M. (1999) *Poststructuralist Geographies: The Diabolical Art of Spatial Science* Edinburgh: Edinburgh University Press

Doel M. and Clarke D. (1994) 'Transpolitical geography' *Geoforum* 25: 505–24

Domosh M. (1998) 'Geography and gender: home, again?' *Progress in Human Geography* 22: 276–82

Donzelot J. (1979) *The Policing of the Family* Paris: Gallimard

Dorling D. and Fairbairn D. (1997) *Mapping: Ways of Representing the World* Harlow: Longman

Douglas M. (1966) *Purity and Danger* Harmondsworth: Penguin

Downs R. (1970) 'Geographic space perception: past approaches and future prospects' *Progress in Geography* 2: 65–108

Driver F. (1988) 'Moral geographies, social science and the urban environment in the mid-nineteenth century' *Transactions, Institute of British Geographers* 13: 275–87

Duncan J. (1993) 'Sites of representation: place, time and the discourse of the other' in Duncan J. and Ley D. (eds) *Place/Culture/Representation* London: Routledge

Duncan J. and Duncan N. (1988) '(Re)reading the landscape' *Environment and Planning D: Society and Space* 6: 117–26

Duncan J. and Ley D. (1993) 'Introduction: representing the place of culture' in Duncan J. and Ley D. (eds) *Place/Culture/Representation* London: Routledge
Duncan N. (1996) 'Renegotiating gender and sexuality in public and private spaces' in Duncan N. (ed.) *Bodyspace* London: Routledge

Eastman C. M. (1975) *Spatial Synthesis in Computer-aided Building Design* London: Applied Science Publishers
Eichberg H. (1998) *Body Cultures: Essays on Sport, Space and Identity* London: Routledge
England K. (1994) 'Getting personal: reflexivity, positionality and feminist research' *The Professional Geographer* 46: 80–89
Entrikin N. (1991) *The Betweenness of Place: Towards a Geography of Modernity* London: Macmillan
Evans D. (1997) *A History of Nature Conservation in Britain* London: Routledge (2nd edn)
Eyles J. (1989) 'The geography of everyday life' in Gregory D. and Walford R. (eds) *Horizons in Human Geography* London: Macmillan

Feyerabend P. (1978) *Against Method* London: Verso
Finkelstein J. (1989) *Dining Out: A Sociology of Modern Manners* Cambridge: Polity Press
Fischer B. and Poland B. (1998) 'Exclusion, risk and social control: reflections on community policing and public health' *Geoforum* 29: 187–97
Flowerdew R. and Martin D. (eds) (1997) *Methods in Human Geography: A Guide for Students doing a Research Project* Harlow: Longman
Foucault M. (1977) *Discipline and Punish* Harmondsworth: Penguin
Foucault M. (1981) *The History of Sexuality* Harmondsworth: Penguin
Foucault M. (1986) 'Of other spaces' *Diacritics* 16: 22–7
Foucault M. (1988) 'The political technology of individuals' in Martin L. H., Gutman P. and Hutton P. H. (eds) *Technologies of the Self* London: Tavistock Press
Franscescato D. and Mebane W. (1973) 'How citizens view two great cities: Milan and Rome' in Downs R. and Stea D. (eds) *Image and Environment: Cognitive Mapping and Spatial Behaviour* Chicago: Aldine Press
Fyfe N. (ed) (1998) *Images of the Street: Planning, Identity and Control in Public Space* London: Routledge
Fyfe N. and Bannister J. (1996) 'City-watching: closed circuit television surveillance in public spaces' *Area* 28: 37–46

Gaarder J. (1996) *Sophie's World* London: Phoenix House
Gallaher C. (1997) 'Identity politics and the religious right: hiding hate in the landscape' *Antipode* 29: 256–77
Gelder K. and Thornton S. (eds) (1997) *The Subcultures Reader* London: Routledge

Giddens A. (1984) *The Constitution of Society* Cambridge: Polity Press
Giddens A. (1990) *The Consequences of Modernity* Cambridge: Polity Press
Gillan A. (1999) 'Abandoned farms will "revert to wilderness"' *Guardian* 1 September: 4
Goffman E. (1963) *Behavior in Public Places: Notes on the Social Organisation of Gatherings* New York: Free Press
Gold J. R. (1980) *An Introduction to Behavioural Geography* Harlow: Longman
Gold J. R. (1982) 'Territoriality and human spatial behaviour' *Progress in Human Geography* 6: 44–67
Gold J. R. (1992) 'Image and environment: the decline of cognitive-behaviouralism in human geography and grounds for regeneration' *Geoforum* 23: 239–47
Gold J. R. and Revill G. (eds) (2000) *Landscapes of Defence* Harlow: Longman
Golledge R. G. (1993) 'Geography and the disabled: a survey with special reference to vision impaired and blind people' *Transactions, Institute of British Geographers* 18: 63–85
Golledge R. G. and Stimson R. J. (1997) *Spatial Behavior: A Geographic Perspective* New York: Guilford Press
Goodey B. and Gold J. R. (1985) 'Behavioural and perceptual geography – from retrospect to prospect' *Progress in Human Geography* 9: 585–95
Graham S., Brooks J. and Heery D. (1996) 'Towns on television: closed circuit TV in British towns and cities' *Local Government Studies* 22: 1–27
Gregory D. (1993) 'The historical geography of modernity: social theory, spatiality and the politics of representation' in Duncan J. and Ley D. (eds) *Place/Culture/Representation* London: Routledge
Gregory D. (1994) *Geographical Imaginations* Oxford: Blackwell
Gregson N. and Lowe M. (1995) 'Home-making: on the spatiality of daily social reproduction in contemporary middle-class Britain' *Transactions, Institute of British Geographers* 20: 224–35
Grosz E. (1994) 'Bodies–cities' in Colomina B. (ed) *Sexuality and Space* Princeton, NJ: Princeton University Press
Guardian (2000) '$350bn media merger heralds net revolution' 11 January: 1

Habermas J. (1991) *Communication and the Evolution of Society* Cambridge: Polity Press
Hägerstrand T. (1982) 'Diorama, path and project' *Tijdschrift voor Economische en Sociale Geografie* 73: 323–39
Haggett P. (1990) *The Geographer's Art* Oxford: Blackwell
Halfacree K. and Kitchin R. (1996) 'Madchester rave on: placing the fragments of popular music' *Area* 28: 47–55
Hall P. (1988) *Cities of Tomorrow* Oxford: Blackwell
Hall P. and Ward C. (1998) *Sociable Cities: The Legacy of Ebenezer Howard* Chichester: John Wiley
Hall S. (1995) 'New cultures for old' in Massey D. and Jess P. (eds) *A Place in the World* Oxford: Oxford University Press/Open University

Hall S. (1997) 'The work of representation' in Hall S. (ed) *Representation: Cultural Representations and Signifying Practices* London: Sage
Halliday S. (1999) *The Great Stink of London* Gloucester: Sutton
Hannah M. (1997) 'Imperfect panopticism' in Benko G. and Strohmayer U. (eds) *Space and Social Theory* Oxford: Blackwell
Haraway D. (1991) *Simians, Cyborgs and Women: The Reinvention of Nature* London: Routledge
Harley J. (1988) 'Maps, knowledge and power' in Cosgrove D. and Daniels S. (eds) *The Iconography of Landscape* Cambridge: Cambridge University Press
Harvey D. (1973) *Social Justice and the City* London: Edward Arnold
Harvey D. (1989) *The Condition of Postmodernity* Oxford: Blackwell
Harvey D. (1992) 'Postmodern morality plays' *Antipode* 24: 300-26
Harvey D. (1993) 'From space to place and back again: reflections on the condition of postmodernity' in Bird J., Curtis B., Putnam T., Robertson G. and Tickner L. (eds) *Mapping the Futures: Local Cultures, Global Change* London: Routledge
Hay A. (1995) 'Concepts of equity, fairness and justice in geographical studies' *Transactions, Institute of British Geographers* 20: 500–8
Hebdidge D. (1979) *Resistance through Rituals* London: Routledge
Herbert D. T. (1993) 'Neighbourhood incivilities and the study of crime in place' *Area* 25: 45–54
Herbert S. (1997) *Policing Space* Minneapolis: University of Minnesota Press
Herman E. S. and McChesney R. W. (1997) *The Global Media: New Missionaries of Corporate Capitalism* London: Cassell
Hetherington K. (1997) *The Badlands of Modernity* London: Routledge
Hewison R. (1987) *The Heritage Industry: Britain in a Climate of Decline* London: Methuen
Hobsbawm E. (1994) *Age of Extremes: The Short Twentieth Century, 1914–1991* London: Michael Joseph
Hofstadter R (1955) *The Age of Reform* New York: Random House
hooks b. (1991) *Yearning: Race, Gender and Cultural Politics* Boston: South End Press
Howard E. (1898) *Tomorrow! A Peaceful Path to Real Reform* London: Swan Sonnenschein
Howard E. (1902) *Garden Cities of Tomorrow* London: Swan Sonnenschein
Hubbard P. J. (1995) 'Urban design in local economic development: a case study of Birmingham' *Cities* 12: 243–53
Hubbard P. J. (1999) *Sex and the City: Geographies of Prostitution in the Urban West* London: Ashgate
Hubbard P. J. (2000) 'Desire/disgust: mapping the moral contours of heterosexuality' *Progress in Human Geography* 24: 191–217
Hudson R. (1995) 'Making music work? Alternative regeneration strategies in a deindustrialized locality: the case of Derwentside' *Transactions, Institute of British Geographers* 20: 460–73

Hughes A., Morris C. and Seymour S. (2000) *Ethnography and Rural Research* Cheltenham: Countryside & Community Press in collaboration with the Rural Geography Study Group

Hull J. (1990) *Touching the Rock: An Experience of Blindness* London: SPCK

Illich I. (1986) *H_2O and the Waters of Forgetfulness* London: Bayers

Imrie R. (1996) 'Ableist geographers, disablist spaces' *Transactions, Institute of British Geographers* 21: 397–403

Ingham J., Purvis M. and Clarke D. B. (1999) 'Hearing places, making spaces: sonorous geographies, ephemeral rhythms and the Blackburn warehouse parties' *Environment and Planning D: Society and Space* 17: 283–305

Jackson P. (1989) *Maps of Meaning* London: Routledge

Jacobs J. (1961) *The Life and Death of Great American Cities* London: Penguin

Jacobs J. M. (1994) 'The battle for Bank Junction: the contested iconography of capital' in Corbridge S., Martin R. and Thrift N. (eds) *Money, Power and Space* Oxford: Blackwell

James S. (1990) 'Is there a "place" for children in human geography?' *Area* 22: 378–83

Jameson F. (1984) 'Postmodernism, or the cultural logic of late capitalism' *New Left Review* 146: 53–92

Janelle D. (1969) 'Spatial reorganisation: a model and concept' *Annals, Association of American Geographers* 59: 336–48

Jervis J. (1999) *Transgressing the Modern: Explorations in the Western Experience of Otherness* Oxford: Blackwell

Johnston R. J. (1991) *Geography and Geographers: Anglo-American Human Geography since 1945* London: Edward Arnold (4th edn)

Johnston R. J. (1997) *Geography and Geographers: Anglo-American Human Geography since 1945* London: Edward Arnold (5th edn)

Johnston R. J., Taylor P. J. and Watts M. J. (eds) (1995) *Geographies of Global Change* Oxford: Blackwell

Katz C. (1994) 'Playing the field: questions of fieldwork in human geography' *The Professional Geographer* 46: 67–72

Kearns G. and Philo C. (eds) (1994) *Selling Places* Oxford: Pergamon Press

Kinsman P. (1995) 'Landscape, race and national identity: the photography of Ingrid Pollard' *Area* 27: 300–10

Kirk W. (1963) 'Problems of geography' *Geography* 48: 357–71

Kitchin R. (1994) 'Cognitive maps: what are they and why study them?' *Journal of Environmental Psychology* 14: 1–19

Kitchin R. (1997) *Geographies of Cyberspace* Chichester: John Wiley

Kitchin R. (1998) 'Towards geographies of cyberspace' *Progress in Human Geography* 22: 385–406

Kitchin R. and Hubbard P. J. (1999) 'Research, action and critical geographies' *Area* 31: 195–8

Kitchin R. and Tate N. J. (2000) *Conducting Research into Human Geography* London: Prentice Hall

Knopp L. (1995) 'Sexuality and urban space: a framework for analysis' in Bell D. and Valentine G. (eds) *Mapping Desire: Geographies of Sexualities* London: Routledge

Kong L. (1990) 'Geography and religion' *Progress in Human Geography* 14: 355–71

Kong L. (1995) 'Music and cultural politics: ideology and resistance in Singapore' *Transactions, Institute of British Geographers* 20: 447–59

Kowinski W. S. (1982) *The Malling of America: An Inside Look at the Great Consumer Paradise* New York: William Morrow

---

Lash S. (1999) *Another Modernity, a Different Rationality* Oxford: Blackwell

Laws G. (1997) 'Women's life courses, spatial mobility and state policies' in Paul-Jones J., Nast H. J. and Roberts S. M. (eds) *Thresholds in Feminist Geography: Difference, Methodology and Representation* New York: Rowman & Littlefield

Lee L. (1959) *Cider with Rosie* London: Hogarth Press

Lefebvre H. (1991) *Critique of Everyday Life* London: Verso

Lewis P. (1985) 'Beyond description' *Annals, Association of American Geographers* 75: 465–77

Ley D. (1980) *Geography Without Man: A Humanistic Critique* Oxford University, School of Geography Research Paper 24 (excerpted under the title 'Geography without human agency: a humanistic critique' in Agnew J., Livingstone D. and Rogers A. (eds) (1996) *Human Geography: An Essential Anthology* Oxford: Blackwell)

Ley D. (1981) 'Behavioural geography and philosophies of meaning' in Cox K. and Golledge R. G. (eds) *Behavioural Problems in Geography Revisited* London: Methuen

Ley D. (1993a) 'Behavioural geography' in Johnston R. J., Gregory D. and Smith D. (eds) *Dictionary of Human Geography* London: Edward Arnold

Ley D. (1993b) 'Postmodernism or the cultural logic of advanced intellectual capital' *Tijdschrift voor Economische en Sociale Geografie* 84: 171–4

Ley D. and Cybriwsky R. (1974) 'Urban graffiti as territorial markers' *Annals, Association of American Geographers* 64: 490–505

Ley D. and Samuels M. (1978) *Humanistic Geography* London: Croom Helm

Leyshon A. (1995) 'Annihilating space? The speed-up of communications' in Allen J. and Hamnett C. (eds) *A Shrinking World?* Oxford: Open University

Leyshon A., Matless D. and Revill G. (1995) 'The place of music' *Transactions, Institute of British Geographers* 20: 423–33

Leyshon A., Matless D. and Revill G. (1998) *The Place of Music* Guildford: Guildford Press

Leyshon A. and Thrift N. (1994) 'A phantom state? The detraditionalization of money, the international financial system and international financial centres' *Political Geography* 13: 299–327
Livingstone D. (1992) *The Geographical Tradition* Oxford: Blackwell
Livingstone D. (1998) 'Reproduction, representation and authenticity: a re-reading' *Transactions, Institute of British Geographers* 23: 13–19
Longhurst R. (1998) 'Representing shopping centres and bodies: questions of pregnancy' in Ainley R. (ed) *New Frontiers of Space, Bodies and Gender* London, Routledge
Lupton D. (1998) *The Emotional Self* London: Sage
Lurie A. (1992) *The Language of Clothes* London: Bloomsbury
Lynch K. (1960) *The Image of the City* Cambridge, MA: MIT Press
Lynch K. (1972) *What Time is this Place?* Cambridge, MA: MIT Press
Lyotard J.-F. (1984) *The Postmodern Condition: A Report on Knowledge* Manchester: University of Manchester Press

McDowell L. (ed.) (1997) *Undoing Place? A Geographical Reader* London: Edward Arnold
McDowell L. (1999) *Gender, Identity and Place: Understanding Feminist Geographies* Cambridge: Polity Press
MacEwen A. and MacEwen M. (1987) *Greenprints for the Countryside: The Story of Britain's National Parks* London: Allen & Unwin
McKay G. (1996) *Senseless Acts of Beauty: Cultures of Resistance since the 1960s* London: Verso
McKendrick J. (1999) 'Playgrounds in the built environment' *Built Environment* 25: 5–11
McLuhen M. (1964) *Understanding Media: Extensions of Man* London: Routledge & Kegan Paul
McNeill D. (1998) 'Writing the new Barcelona' in Hall T. and Hubbard P. J. (eds) *The Entrepreneurial City: Geographies of Politics, Regime and Representation* Chichester: John Wiley
Maffesoli M. (1989) *The Time of the Tribes* London: Sage
Markus T. (1993) *Buildings and Power: Freedom and Control in the Evolution of Modern Building Types* London: Routledge
Marsh J. (1982) *Back to the Land* London: Quartet
Martin B. (1984) 'Mother wouldn't like it! Housework as magic' *Theory, Culture and Society* 2: 1–28
Massey D. (1991) 'The political place of locality studies' *Environment and Planning A* 23: 267–81
Massey D. (1993) 'Power geometry and a progressive sense of place' in Bird J., Curtis T., Putnam T., Robertson G. and Tickner L. (eds) *Mapping the Futures* London: Routledge
Massey D. (1995a) 'The conceptualisation of place' in Massey D. and Jess P. (eds) *A Place in the World?* Oxford: Open University Press

Massey D. (1995b) 'Imagining the world' in Allen J. and Massey D. (eds) *Geographical Worlds* Oxford: Open University Press
Matless D. (1990) 'Definitions of England, 1928–89: preservation, Modernism and the nature of the nation' *Built Environment* 16: 179–91
Matless D. (1993) 'Appropriate geography: Patrick Abercrombie and the energy of the world' *Journal of Design History* 6: 167–78
Matless D. (1995) 'The art of right living' in Pile S. and Thrift N. (eds) *Mapping the Subject: Geographies of Cultural Transformation* London: Routledge
Matless D. (1998) *Landscape and Englishness* London: Reaktion Press
Matthews M. H. (1992) *Making Sense of Place: Children's Understanding of Large-scale Environments* Henel Hempstead: Harvester Wheatsheaf
Matthews M. H. (1995) 'Living on the edge: children as outsiders' *Tijdschrift voor Economische en Social Geografie* 86: 456–66
Matthews M. H. and Vujakovic P. (1995) 'Private worlds and public places: mapping the environmental values of wheelchair users' *Environment and Planning A* 27: 1069–83
May J. (1996) 'Globalization and the politics of place: place and identity in an inner London neighbourhood' *Transactions, Institute of British Geographers* 21: 194–215
Merrifield A. and Swyngedouw E. (eds) (1996) *The Urbanisation of Injustice* London: Lawrence & Wishart
Milbourne P. (ed) (1997) *Revealing Rural 'Others'* London: Cassell
Miles M. (1997) *Art, Space and the City* London: Routledge
Miller T. and McHoul A. (1998) *Popular Culture and Everyday Life* London: Sage
Mingay G. (ed) (1989) *The Rural Idyll* London: Routledge
Mitchell D. (1997) 'The annihilation of space by law: the roots and implications of anti-homeless laws in the United States' *Antipode* 29: 303–35
Mitchell W. J. (1996) *City of Bits* Chicago: MIT Press
Mooney G. (1999) 'Urban disorders' in Pile S., Brook C. and Mooney G. (eds) *Unruly Cities: Order/Disorder* London: Routledge
Mordue T. (1999) 'Heartbeat country: conflicting values, coinciding visions' *Environment and Planning A* 31: 629–46
Murdoch J. and Pratt A. (1993) 'Rural studies: modernism, postmodernism and the post-rural' *Journal of Rural Studies* 9: 411–27
Murdoch J. (1997) 'Towards a geography of heterogeneous associations' *Progress in Human Geography* 21: 321–7
Murdoch J. (1998) 'The spaces of actor-network theory' *Geoforum* 29: 357–74
Myers G. (1998) *Ad Worlds* London: Edward Arnold

Nash C. (1996) 'Reclaiming vision: looking at landscape and the body' *Gender, Place and Culture* 3: 149–69
Nead L. (1997) 'Mapping the self: gender, space and modernity in mid-Victorian London' *Environment and Planning A* 29: 659–72

Oc T. and Tiesdell S. (eds) (1997) *Safer City Centres: Reviving the Public Realm* London: Paul Chapman

Ogborn M. (1988) 'Love–state–ego: centres and margins in nineteenth-century Britain' *Environment and Planning D: Society and Space* 10: 287–305

Ogborn M. (1998) *Spaces of Modernity: London's Geographies 1680–1780* New York: Guilford Press

Ogborn M. and Philo C. (1994) 'Soldiers, sailors and moral locations in nineteenth-century Portsmouth' *Area* 27: 215–31

Pacione M. (1999) 'Applied geography: in pursuit of useful knowledge' *Applied Geography* 19: 1–12

Pain R. (1991) 'Space, sexual violence and social control: integrating geographical and feminist analyses of women's fear of crime' *Progress in Human Geography* 15: 415–31

Pain R. (1997) 'Social geographies of women's fear of crime' *Transactions, Institute of British Geographers* 22: 231–44

Park R. (1916) 'The city: suggestions for the investigation of human behaviour in the urban environment' *American Journal of Sociology* 20: 577–612

Parkes D. and Thrift N. (1980) *Times, Spaces and Places: A Chronogeographic Perspective* Chichester: John Wiley

Peet R. (1998) *Modern Geographical Thought* Oxford: Blackwell

Pepper D. (1996) *Modern Environmentalism: An Introduction* London: Routledge

Phillips M. (1998) 'Restructuring of social imaginations in rural geography' *Journal of Rural Studies* 14: 121–53

Philo C. (1987) 'Fit localities for an asylum' *Journal of Historical Geography* 13: 398–415

Philo C. (1992) 'Neglected rural geographies: a review' *Journal of Rural Studies* 8: 429–36

Philo C. (1997) 'Of other rurals' in Cloke P. and Little J. (eds) *Contested Countryside Cultures* London: Routledge

Pile S. (1991) 'Practising interpretative geography' *Transactions, Institute of British Geographers* 16: 458–69

Pile S. (1993) 'Human agency and human geography revisited: a critique of "new models" of the self' *Transactions, Institute of British Geographers* 18: 122–39

Pile S. (1996) *The Body and the City* London: Routledge

Pile S. (1999) 'The heterogeneity of the city' in Pile S., Brook C. and Mooney G. (eds) *Unruly Cities: Order/Disorder* London: Routledge

Pile S. and Thrift N. (eds) (1995) *Mapping the Subject: Geographies of Cultural Transformation* London: Routledge

Pile S. and Keith M. (eds) (1997) *Geographies of Resistance* London: Routledge

Pile S. and Nast H. (eds) (1998) *Places through the Body* London: Routledge

Pile S., Brook C. and Mooney G. (eds) (1999) *Unruly Cities: Order/Disorder* London: Routledge

Pocock D. (1993) 'The senses in focus' *Area* 25: 11–16

Pocock D. (1996) 'Place evocation: the Galilee Chapel in Durham Cathedral' *Transactions, Institute of British Geographers* 21: 379–86

Porteous D. (1985) 'Smellscape' *Progress in Human Geography* 9: 356–78

Pratt G. and Hanson S. (1994) 'Geography and the construction of difference' *Gender, Place and Culture* 1: 5–29

Pred A. (1977) 'The choreography of existence' *Economic Geography* 53: 207–21

Pred A. (1985) 'The social becomes the spatial, the spatial becomes the social: enclosures, social change and the becoming of place in Skane' in Gregory D. and Urry J. (eds) *Social Relations and Spatial Structure* London: Macmillan

Pred A. (1990) *Lost Words and Lost Worlds: Modernity and the Language of Everyday Life in Late Nineteenth-century Stockholm* Cambridge: Cambridge University Press

Procter J. (1998) 'Ethics in geography: giving moral form to the geographical imagination' *Area* 30: 8–18

Proshansky H. M. (1976) *Environmental Psychology: People in their Physical Setting* New York: Holt, Rinehart & Winston

Putnam T. (1993) 'Shifting the parameters of residence' in Bird J., Curtis T., Putnam T., Robertson G. and Tickner L. (eds) *Mapping the Futures* London: Routledge

Rapoport A. (1972) 'Australian aborigines and the definition of place' in Mitchell W. (ed.) *Environmental Design: Research and Practice* Los Angeles: Proceedings of the 3rd EDRA conference, 3-3-1 – 3-3-14

Ratcliffe B. M. (1990) 'Cities and environmental design: elites and sewerage problems in Paris from the mid-eighteenth to the mid-nineteenth century' *Planning Perspectives* 5: 189–222

Relph E. (1976) *Place and Placelessness* London: Pion

Relph E. (1987) *The Modern Urban Landscape* London: Pion

Relph E. (1989) 'Responsive methods; geographical information and the study of landscapes' in Kobayashi A. and Mackenzie S. (eds) *Remaking Human Geography* London: Unwin Hyman

Richardson L. (1994) 'Writing: a method of enquiring' in Denzin N. K. and Lincoln Y. S. (eds) *Handbook of Qualitative Research* Thousand Oaks, CA: Sage

Ritzer G. (1993) *The McDonaldization of Society: An Investigation into the Changing Character of Contemporary Social Life* London: Sage

Roberts J. (1999) 'Philosophising the everyday: the philosophy of praxis and the fate of cultural studies' *Radical Philosophy* 98: 16–29

Rodaway P. (1994) *Sensuous Geographies: Body, Sense and Place* London: Routledge

Rose G. (1993) *Feminism and Geography* Cambridge: Cambridge University Press

Rose G. (1995) 'Place and identity: a sense of place' in Massey D. and Jess P. (eds) *A Place in the World* Oxford: Open University Press

Roszak T. (1972) *Where the World Ends: Politics and Transcendence in Post-industrial Society* New York: Doubleday

Said E. (1978) *Orientalism* New York: Pantheon
Savage M. and Warde A. (1993) *Urban Sociology, Capitalism and Modernity* Basingstoke: Macmillan
Schlör J. (1998) *Nights in the Big City* London: Reaktion Press
Schoenberger E. (1998) 'Discourse and practice in human geography' *Progress in Human Geography* 22: 1–14
Seamon D. (1979) *A Geography of the Lifeworld* London: Croom Helm
Semple E. (1911) *Influences of Geographic Environment: On the Basis of Ratzel's System of Anthropo-geography* London: Constable
Sennett R. (1977) *The Fall of Public Man* London: Faber & Faber
Sennett R. (1994) *The Flesh and the Stone: The Body and the City in Western Civilisation* London: Faber & Faber
Seymour J. (1991) *Rural Life: Pictures from the Past* Collins & Brown
Shelton A. (1990) 'A theater for eating, looking and thinking: the restaurant as a symbolic space' *Sociological Spectrum* 10: 507–26
Shields R. (1991) *Places on the Margin* London: Routledge
Shields R. (1996) 'A guide to urban representation and what to do with it' in King A. (ed.) *Representing the City: Ethnicity, Capital and Culture in the 21st Century Metropolis* Cambridge: Polity Press
Shields R. (1997) 'Spatial stress and resistance: social meanings of spatialisation' in Benko G. and Strohmayer U. (eds) *Space and Social Theory* Oxford: Blackwell
Shilling C. (1993) *The Body and Social Theory* London: Sage
Short J. R. (1991) *Imagined Country: Society, Culture and Environment* London: Routledge
Short J. R. (1996) *The Urban Order: An Introduction to Cities, Culture and Power* Oxford: Blackwell
Short J. R. and Kim Y.-H. (1998) 'Urban crises/urban representation' in Hall T. and Hubbard P. (eds) *The Entrepreneurial City: Geographies of Politics, Regime and Representation* Chichester: John Wiley
Short J. R. and Kim Y.-H. (1999) *Globalization and the City* Harlow: Longman
Short J. R. and Lowe M. (1990) 'Progressive human geography' *Progress in Human Geography* 14: 1–11
Shurmer-Smith P. and Hannam K. (1994) *Worlds of Desire Realms of Power,* London: Routledge
Sibley D. (1995) *Geographies of Exclusion: Society and Difference in the Urban West* London: Routledge
Sibley D. (1998) 'The problematic nature of exclusion' *Geoforum* 29: 119–21
Sidaway J. (1992) 'In other worlds: on the politics of research by "first world" geographers in the "third world"' *Area* 24: 403–8
Sidaway J. (1997) 'The production of British geography' *Transactions, Institute of British Geographers* 22: 488–504
Silverstone R. (ed.) (1997) *Visions of Suburbia* London: Routledge
Sinclair I. (1997) *Lights out for the Territory* London: Granta
Skurnik L. and George F. (1967) *Psychology for Everyone* Harmondsworth: Penguin
Smart B. (1993) *Postmodernism* London: Routledge

Smith D. M. (1975) *Human Geography: A Welfare Approach* London: Edward Arnold
Smith D. M. (1995) *Geography and Social Justice* Oxford: Blackwell
Smith D. M. (1997a) 'Geography and ethics: a moral turn?' *Progress in Human Geography* 21: 583–90
Smith D. M. (1997b) 'Back to the good life: towards an enlarged conception of social justice' *Environment and Planning D: Society and Space* 15: 19–35
Smith D. M. (2000) *Moral Geographies: Ethics in a World of Difference* Edinburgh: Edinburgh University Press
Smith N. (1996a) 'Rethinking sleep' *Environment and Planning D: Society and Space* 14: 505–6
Smith N. (1996b) *The New Urban Frontier: Gentrification and the Revanchist City* London: Routledge
Smith S. (1984) 'Crime and the structure of social relations' *Transactions, Institute of British Geographers* 9: 427–42
Smith S. (1999) 'Society–space' in Cloke P., Crang P. and Goodwin M. (eds) *Introducing Human Geographies* London: Edward Arnold
Soja E. (1989) *Postmodern Geographies: The Reassertion of Space in Critical Social Theory* London: Verso
Soja E. (1996) *Thirdspace* Oxford: Blackwell
Soper K. (1995) *What is Nature? Culture, Politics and the Non-human* Oxford: Blackwell
Sommer R. (1969) *Personal Space: The Behavioral Basis of Design* Englewood Cliffs, NJ: Prentice-Hall
Sparke M. (1996) 'Displacing the field in fieldwork: masculinity, metaphor and space' in Duncan N. (ed.) *Bodyspace: Destabilising Geographies of Gender and Sexuality* London: Routledge
Spencer C. (1991) 'Life-span changes in activities and consequent changes in the cognition and assessment of the environment' in Garling T. and Evans G. W. (eds) *Environment, Cognition and Action: An Integrated Approach* Oxford: Oxford University Press
Spooner D. (1992) 'Places I'll remember . . . Larkin's *Here*' *Geography* 77: 134–42
Stallybrass P. and White A. (1986) *The Politics and Poetics of Transgression* London: Methuen

Taylor P. J. (1991) 'The English and their Englishness: "a curiously mysterious, elusive and little understood people"' *Scottish Geographical Magazine* 107: 146–61
Taylor P. J. (1999) 'A geohistorical interpretation of the modern world' in Cloke P., Crang P. and Goodwin M. (eds) *Introducing Human Geographies* London: Edward Arnold
Teather E. M. (1999) *Embodied Geographies: Space, Bodies and Rites of Passage* London: Routledge
Tester K. (1992) *Civil Society* London: Routledge
Thomas D. (1946) *Deaths and Entrances* London: J. M. Dent
Thomas K. (1983) *Man and the Natural World: Changing Attitudes in England, 1500–1800* London: Allen Lane

Thompson D. L. (1963) 'New concept: subjective distance' *Journal of Retailing* 39: 1–6
Thompson J. (1990) *Ideology and Modern Culture* Cambridge: Cambridge University Press
Thrift N. (1994) 'Inhuman geographies: landscapes of speed, light, power' in Cloke P. J., Doel M. A., Matless D., Phillips M. and Thrift N. (eds) *Writing the Rural* London: Paul Chapman
Thrift N. (1995) 'A hyperactive world' in Johnston R. J., Taylor P. J. and Watts M. J. (eds) *Geographies of Global Change* Oxford: Blackwell
Thrift N. (1997) 'Us and them: re-imaging places, re-imaging identities' in Mackay H. (ed) *Consumption and Everyday Life* London: Sage
Tomlinson J. (1997) 'Internationalisation, globalisation and cultural imperialism' in Thompson K. (ed) *Media and Cultural Regulation* London: Sage
Tuan Y.-F. (1974) 'Space and place: a humanistic perspective' *Progress in Geography* 6: 233–46 (excerpted in Agnew J., Livingstone D. and Rogers A. (eds) (1996) *Human Geography: An Essential Anthology* Oxford: Blackwell)
Tuan Y.-F. (1977) *Space and Place: The Perspective of Experience* Minneapolis: University Minnesota Press
Tuan Y.-F. (1978) 'Literature and geography: implications for geographical research' in Ley D. and Samuels I. (eds) *Humanistic Geography* London: Croom Helm
Tuan Y.-F. (1979) *Landscapes of Fear* Oxford: Blackwell
Turner B. (1988) *Status* Milton Keynes: Open University Press

Unwin T. (1992) *The Place of Geography* London: Unwin Hyman
Urry J. (1990) *The Tourist Gaze* London: Sage

Valentine G. (1989) 'The geography of women's fear' *Area* 21: 385–90
Valentine G. (1992) 'Images of danger: women's sources of information about the spatial distribution of male violence' *Area* 24: 22–9
Valentine G. (1993) 'Hetero-sexing space: lesbian perceptions and experiences of everyday spaces' *Environment and Planning D: Society and Space* 9: 395–413
Valentine G. (1995) 'Creating transgressive space: the music of kd lang' *Transactions, Institute of British Geographers* 20: 474–86
Valentine G. (1996) 'Children should be seen and not heard: the production and transgression of adults' public space' *Urban Geography* 17: 205–20
Valentine G. (1998) 'Food and the production of the civilised street' in Fyfe N. (ed) *Images of the Street: Planning, Identity and Control in Public Space* London: Routledge
Valentine G. (1999) 'A corporeal geography of consumption' *Environment and Planning D: Society and Space* 17: 329–51

Walmsley D. J. (1988) *Urban Living: The Individual in the City* Harlow: Longman
Walmsley D. J. and Lewis G. (1993) *People and Environment: Behavioural Approaches in Human Geography* Harlow: Longman

Ward C. (1990a) *The Child in the City* London: Architectural Press (2nd edn)
Ward C. (1990b) *The Child in the Country* London: Bedford Square Press (2nd edn)
Ward L. (1999) 'Prescott to unveil new national parks' *Guardian* 29 September: 7
Warf B. (1993) 'Postmodernism and the localities debate: ontological questions and epistemological implications' *Tijdschrift voor Economische en Sociale Geografie* 84: 162–9
Watson S. (1986) 'Housing the family: the marginalisation of non-family households in Britain' *International Journal of Urban and Regional Research* 10: 8–28
Watson S. (1991) 'Gilding the smokestacks: new symbolic representations of deindustrialised regions' *Environment and Planning D: Society and Space* 9: 59–70
Watson S. (1999) 'City politics' in Pile S., Brook C. and Mooney G. (eds) *Unruly Cities: Order/Disorder* London: Routledge
Watts M. (1991) 'Mapping meaning, denoting difference, imagining identity: dialectic images and postmodern geographies' *Geografiska Annaler* 73B: 7–16
Weinstein C. S. (1979) 'The physical environment of the school: a review of research' *Review of Educational Research* 49: 577–610
Whatmore S. (1999) 'Culture–nature' in Cloke P., Crang P. and Goodwin M. (eds) (1999) *Introducing Human Geographies* London: Arnold
Whatmore S. and Thorne L. (1998) 'Wild(er)ness: reconfiguring the geographies of wildlife' *Transactions, Institute of British Geographers* 23: 435–54
Wiener M. (1981) *English Culture and the Decline of the Industrial Spirit* Harmondsworth: Penguin
Wilton R. (1996) 'Diminished worlds? The geography of everyday life with HIV/AIDS' *Health and Place* 2: 69–83
Williams R. (1973) *The Country and the City* London: Chatto & Windus
Winchester H. and White P. (1988) 'The location of marginalised groups in the inner city' *Environment and Planning D: Society and Space* 6: 37–54
Winter M. (1996) *Rural Politics: Policies for Agriculture, Forestry and the Environment* London: Routledge
Wolch J. and Emel J. (eds) (1998) *Animal Geographies: Place, Politics and Identity in the Nature–Culture Borderlands* London: Verso
Women and Geography Study Group (1997) *Feminist Geographies: Explorations in Diversity and Difference* Harlow: Longman
Wood D. (1993) *The Power of Maps* London: Routledge
Woods V. (1999) 'It's not the sexy undies, its the unsexy lighting' *Sunday Telegraph* 23 May: 23
Wynne B. (1996) 'May the sheep safely graze? A reflexive view of the expert–lay knowledge divide' in Lash S., Szerszynski B. and Wynne B. (eds) *Risk, Environment and Modernity: Towards a New Ecology* London: Sage

Young I. M. (1990) *Justice and the Politics of Difference* Princeton, NJ: Princeton University Press

Zukin S. (1995) *The Culture of Cities* Oxford: Blackwell

index